MW00721549

# FLYING CANUCKS
## Famous Canadian Aviators

*For my father-in-law, G.S. O'Neill,*
*who flew with the Canucks in 1944-45.*

# FLYING CANUCKS
## Famous Canadian Aviators

## PETER PIGOTT

HOUNSLOW

Flying Canucks: Famous Canadian Aviators

**Hounslow Press**
A member of the Dundurn Group

Publishers: Kirk Howard & Anthony Hawke
Editor: Dennis Mills
Printer: Metrolitho Inc., Quebec
Interior photographs: National Aviation Museum and the Public Archives of Canada
Front Cover Painting:  Bellanca CH - 300 Pacemaker (General Airways Limited, 1933), by Robert W. Bradford

**Canadian Cataloguing in Publication Data**

Pigott, Peter
   Flying Canucks

ISBN 0- 88882-175-1

1. Air pilots - Canada - Biography. 2. Aeronautics -
Canada - History. I. Title.

TL539.P54 1994     629. 13'0092'271    C94-932627-5

Publication was assisted by the **Canada Council**, the **Book Publishing Industry Development Program** of the **Department of Canadian Heritage**, the **Ontario Arts Council**, and the Ontario Publishing Centre of the **Ontario Ministry of Culture, Tourism and Recreation.**

Care has been taken to trace the ownership of copyright material used in this book. The author and the publisher welcome any information enabling them to rectify any references or credit in subsequent editions.

Printed and bound in Canada

| Hounslow Press | Hounslow Press | Hounslow Press |
|---|---|---|
| 2181 Queen Street East | 73 Lime Walk | 1823 Maryland Avenue |
| Suite 301 | Headington, Oxford | P.O. Box 1000 |
| Toronto, Canada | England | Niagara Falls, NY |
| M4E 1E5 | OX3 7AD | U.S.A. 14302-1000 |

# CONTENTS

# ACKNOWLEDGEMENTS

The subject of this book is close to my heart and I am indebted to the people who made its writing possible. After more than a decade overseas with the Department of Foreign Affairs, I was shipwrecked in Ottawa. Fortunately, I washed ashore at the Bureau of Assistance for Central & Eastern Europe where my colleagues, in spite of heavy commitments, gave me their encouragement and understanding. Like the subjects of this work, they exemplify the very best in the Canadian spirit.

I am also indebted to the staff at the Lester B. Pearson Library and to Claudette St. Hilaire at the National Aviation Museum.

# INTRODUCTION

For most of its existence, this country has been identified with one form of transportation or another. The birchbark canoe, the steam locomotive, and at the dawn of this century, flimsy contraptions held together with wire and muslin ... all were pivotal to our national identity.

This book is not a comprehensive collection of biographies of Canadians who pioneered aviation or aspired to do so. It focuses on a certain phase in our history — barely one lifetime — that of the evolution of aviation in Canada. On one level, the men and women chosen took a piece of applied science and changed our country forever.

It has not been possible, even with the most diligent searching, to document fully the lives of several aviators and it would be futile to analyze why they embarked on the courses that they did. By themselves the biographies make little sense, and it has been necessary to connect them with the events of their day — the aircraft used, the aviation policies and organizations.

Another complication is that while this work is about the development of aviation in Canada, especially in the early years of this century, few of its pioneers were Canadian born. The very term *Canadian national* was not legally recognized until legislation in 1921; prior to that date, persons living in this country were British subjects. Also, there were many native-born Canadians who served with the British forces during and after the two world wars. As they were undoubtedly Canadian when they performed their aerial feats of distinction, they have been included.

The admission of Americans like Glenn Curtiss, James Dalzell McKee, and Katherine Stinson deserves some explanation, but no apology. Canadian aviation was fortunate to evolve beside the most air-minded nation in the world. At a time when the distinction between national achievements was blurred, when our tiny population and vast distances put major technological advances beyond the slender means of either Ottawa or Bay Street, the spill-over of American vision and resources was welcome. It gave an impetus to our national dream.

For those who look in vain for R.W. 'Buck' McNair, Frank Young and Barney Rawson, my apologies. The choice of aviators in this book is purely a personal one, based on the availability of primary sources. That Alexander Graham Bell is mentioned only in connection with the Aerial Experimental Association deserves an explanation. While the AEA had been financed by his wife, Bell's brilliance encompassed myriad inventions like the telephone, the gramophone and the hydrofoil. But once his tetrahedral kites *Frost King* and *Cygnet* had satisfied his curiosity of aeronautics, like a lot of inventors, he moved on to other interests. Besides, he was now in his seventies and quite content to leave the field to younger men.

Finally, incorporating into this history a few remarkable fliers who were awarded the highest accolade for their bravery in the air, does not distort it. Their self sacrifice — in the Second World War all Victoria Crosses were awarded to Canadian airmen posthumously — publicised how far the country had come. No longer struggling to get out from under Britain's shadow, Canadians believed that they could do anything.

Carl Agar in a Bell 47D Helicopter. (NAM 3406)

# AGAR, CARL C.

T o write of Carl Agar is to write of flying helicopters in mountainous terrain. Even while this mode of flying was in its infancy, Agar pioneered the use of helicopters at high altitude, opening a new era for both the machine and previously inaccessible construction projects around the world.

Carl Clare Agar was born at Lion's Head, Ontario, on November 28, 1901. His brother Egan lost his life in the Royal Flying Corps in the First World War, but became interested in aviation only after meeting the legendary 'Wop' May, and in 1929 joined the Edmonton and Northern Alberta Aero Flying Club. After obtaining his own private pilot's licence, Agar farmed until the outbreak of the Second World War. Like many famous pilots in post-war Canadian aviation, Agar served as an instructor in the British Commonwealth Air Training Plan throughout the war. Because of his age, he was refused an overseas posting and worked in Abbotsford, British Columbia, at the Elementary Flight Training School.

In 1945 he was demobilised and entered commercial aviation; with other veterans he formed the Okanagan Air Service at Kelowna, British Columbia. Like many ventures launched in the post-war years, the OAS would have gone bankrupt had the partners not driven to Washington State in October, 1946. There they saw a revolutionary crop-spraying aircraft. It was a Bell 47 B3 helicopter, the original bubbled-shaped dragon fly, the first Agar had seen.

With the support of farmers in the Okanagan valley, the partners purchased a helicopter to spray and dust local orchards, and Agar took flying lessons at Yakima, Washington. While this was a start, OAS looked for other business during the lean months. It was then that he had the idea of using the fragile machine as "a pack horse of the air." Operating it at high altitudes in the mountains was then a wholly radical concept, and he was forced to adapt and invent techniques that were different from flying over water or flat terrain. He had been warned that helicopters would not work over 4,500 feet, and that the thin atmosphere in the Rockies would prevent a Bell from gaining full take-off power. The neophyte pilot had been told that the updrafts and downdrafts in the mountains were fickle and that in any case there was no space on the top of a mountain to land a helicopter.

With the determination that characterized him all his life, Agar worked for years to become an expert helicopter pilot. Wartime flying at Abbottsford had given him some of the expertise he needed. But on one occasion it looked like the critics were right when they said that a helicopter could not operate at high altitudes. Agar was flying over the Coastal Range and landed on a clifftop, at 7,500 feet. As the thin atmosphere prevented a conventional take-off, he thought he might be stuck forever. He used full power to get the machine a few feet off the ground and then literally dumped it over the cliff-side. It fell about 500 feet before the blades could take hold.

In the late '40s his company was becoming well known as the authority on mountain helicopter flying. In 1949, the forestry branch of the British Columbia government gave them the contract to spray against infestations in the Fraser Valley. The Greater Vancouver Water District later asked if OAS could move all the stores and equipment to build their dam on Palisade Lake. This meant that 400,000 pounds of material would have to be lifted to a spot 3,000 feet above sea level. When one considers that the Bell 47 could carry a mere 400 pounds per trip, the problems were phenomenal. The operation required thousands of flights; steel, lumber, cement, concrete mixers were carried high into the mountains. On one memorable day, the hard-working Bell made 41 trips up the mountain. The successful completion of the dam, the first in the world to be built with the use of helicopters, was all the proof that Agar needed about the versatility of the machine.

Prospecting by air was not new, but the mining companies now saw that they could investigate more closely the mineral wealth locked in the mountain ranges. The ruggedness of the Cheam Range in British Columbia had made it uneconomical to locate survey camps overland. With their machines, OAS carried men and supplies in an hour to co-ordinates on a map that would have taken days by land. Heavy equipment, wood for whole bunkhouses, and all the miners were airlifted by the helicopters.

Like the bush pilots of the 1920s, Agar had gas caches in half-gallon drums hauled to the summits of mountains so that his helicopters did not have to return to base every night. These "fly camps" meant that several prospecting parties could be supplied at the same time, the helicopter rotating from one to the other. Flight operations at 6,000-foot levels became common-place.

In 1952 OAS purchased the larger S-55s to work on the big Alcan alu-minium factory construction project at Kemano River, British Columbia. Soon the S-58 was added, and by 1958, OAS owned 21 S-55s and 31 S-58s, with most of the latter working on the DEW Line.

Agar was friendly with the early Canadian helicopter designer Bernard W. Sznycer. This Polish emigré had built a small helicopter plant in Montreal in 1946 and tried to interest OAS in his "Omega" model. With its twin engines and ability to fly on either engine, the Omega would have been certified to operate over cities, and Agar considered going into the passenger ferry business.

Unfortunately for both Agar and Sznycer, the financiers backed out of the project before the machine could be put into production. A chance to compete with the giant Sikorsky company was irrevocably lost to Canada.

OAS continued to expand their operations with 'timber cruising' and running a flying training school. On several occasions, Agar was called in by foreign governments to give advice on the use of helicopters in adverse conditions. From the jungles of New Guinea to the Himalayas, Canadian expertise in vertical flight was being sought. In the Western Arctic islands, surveying was conducted with Okanagan machines for the Army Survey Establishment. In 1956, Agar took over rival Canadian Helicopters and expanded overseas on a unprecedented scale. In 1977 OAS had five S-61s on off-shore oil exploration work around the world. There was one each in Halifax; an Alaskan island; Shannon, Erie; Bombay, India; and Ivercagill, New Zealand. Its latest purchase, the giant S-61 with a seating capacity of 28 passengers, was an airliner in its own right, a far cry from the early, bubble-shaped Bell 47s.

In 1962 due to ill health Carl Agar gave up his position in Okanagan, only to rejoin it in 1965. Sadly, "Mr. Helicopter," as he had become known in the aviation world, died of a lung disorder on January 26, 1968.

Of him, the inventor of the helicopter, Igor Sikorsky said: "A flying craft remains useless unless there also exists pioneers with courage, foresight and energy, who can visualize the usefulness of it and fulfil the final stage of its development ... to prove its value and thus assign to it its rightful place in our modern life. Carl Agar, to my mind is one of the most brilliant and outstanding pioneers of this type." [1]

---

1 Milberry, L., *Aviation in Canada* (Toronto: McGraw-Hill Ryerson, 1979), p. 244.

Russell Bannock. He attributed much of his success as a night fighter ace to his instructing experience in the BCATP. (NAM11552)

# BANNOCK, RUSSELL

If, as an instructor in the British Commonwealth Air Training Plan, Russell Bannock had few equals, as a night-fighter pilot against the German V bombs, he was unequalled. A former bush pilot, he was instrumental in selling worldwide that favourite aircraft of modern  bushpilots — the de Havilland Beaver.

Born at Edmonton, Alberta on November 1, 1919, Russell Bannock became infatuated with aviation at an early age. As a boy he spent many happy hours at Edmonton's Blatchford Airport in the heady 1920s. Having seen the great bush pilots like 'Punch" Dickins and 'Wop' May fly in and out, or Americans like Wiley Post and Wallace Beery refuel on their way to Alaska, who could not be impressed? But careers in aviation during the Depression were a gamble, and Bannock settled on becoming a geologist.

He attended the University of Calgary and began a routine familiar to university students — of working during the summer to finance his university education. His first flight took place in 1937 when he flew to Yellowknife with the aviation pioneer Stan McMillan to take a summer job as bar steward for the Hudson's Bay Company. Bannock always remembers that due to poor weather that day, they never flew above a frightening hundred feet. But he was hooked, and that autumn he began taking flying lessons. His first full-time job was working for Grant McConachie's Yukon Southern Transport as a co-pilot.

At the outbreak of war, as a trained pilot, Bannock was immediately recruited by the Ministry of Defence into the RCAF's 112 Squadron. This was an Army Co-operation squadron using Lysanders, and the young pilot was about to leave for France and the British Expeditionary Force, when events overtook his troop ship in Halifax harbour. The suddenness of the German advance and the Allied retreat at Dunkirk caused all pilots to be recalled and made into the nucleus of the British Commonwealth Air Training Plan (BCATP).

Chaffing to get into the air battles that were now taking place over England, Bannock was instead forced to settle for the role of a flight instructor at Trenton, Ontario. By September 1942, he was promoted to the Chief Instructor of No. 3 Flying School at Arnprior, Ontario.

His request for an overseas posting was finally granted in 1944 and he found himself at No. 60 Operational Training Unit at High Ercall, Shropshire, England, with the 418 Squadron RCAF.

The squadron had been flying Douglas DB-7 Bostons on 'intruder' operations. Their role was to penetrate deep into enemy territory alone at night and shoot down enemy night fighters as they took off or landed. In danger of being shot down by the enemy as well being hit by their own flak or barrage balloons, the night- fighter crews, on both sides, operated entirely by their wits, relying only on each other and their machines. The RCAF's Bostons had not been designed for this role and were unwieldy to fly, but fortunately 418 Squadron exchanged them for the inimitable Mosquitoes in 1942.

Made of wood by de Havilland - in what was called the 'glue-and-screw' method of manufacturing - the Mosquito was fast and manoeuvrable. It was also heavily armed with four 20 mm Hispano cannons mounted underneath the aircraft and four .303 machine guns in its nose. It had been conceived as an unarmed bomber that would rely on its speed for defence, and at 20,000 feet it was capable of 400 mph, faster than most of the fighters of its day.

Now based at Arundel, Surrey, the Canadians began supporting Bomber Command in what were known as 'Flower Operations'. This meant covering German airfields during a massed Allied night raid on a strategic target. Bannock and his colleagues would position their Mosquitoes over an enemy night-fighter home base and wait for them to land or take off. The Luftwaffe crews were forced to burn their navigation lights at these vulnerable times, if only to prevent collisions and not be shot at by their own anti-aircraft units. This meant that for a few seconds they became illuminated targets for the 'intruder' lying in ambush, and 418 took full advantage to inflict heavy causalities.

With the ranks of his air force severely depleted of trained pilots, in 1944 Hitler resorted to using V-1 flying bombs. Prime Minister Churchill feared the demoralizing consequences of a massive attack against London, and he personally ordered that all Mosquito squadrons operate over the Channel on anti-V-1 patrols.

The V-1s were pilotless aircraft, about 20 feet in length, with wooden wings and a metal fuselage. They had a range of 375 miles and carried a 1,500-pound warhead. The 'buzz bombs' were a cheap means of terror bombing, launched in packs of 20 or 30 to overwhelm Allied defences.

On June 16, 1944, while the Allied armies were securing their foothold in Normandy, W/C Bannock on night patrol saw what he thought was a burning aircraft crossing the English Channel. It was one of the first of the V-1s.

With other night-fighter squadrons, 418 soon found out that the V-1s were faster than their aircraft and that the only time a Mosquito could match their speed was in a dive. Accordingly, a whole strategy was developed.

The night fighter would hover over the Channel and wait for a launch of a V-1. As soon as the crew saw one fired, they would turn their Mosquito around and speed towards London keeping ahead of it, always looking over their shoulders as it caught up. When the buzz-bomb was directly under, the pilot would switch on the navigation lights to warn other aircraft, and dive steeply upon it.

Bannock recalled in a speech many years later, that it took all his strength to keep the light Mosquito in a dive at full rudder because of the torque effect. Very aware of the warhead on board, the crew had to fire on a 'buzz bomb' at no less than 300 yards, to avoid being caught in the explosion. But the German night fighters soon saw the defence strategy, and in their own version of 'Flower Operations', began to pick off the illuminated Mosquitos as they dived onto the V-1s.

W/C Bannock and his navigator F/O Robert Bruce were deeply committed in the flying bomb battle. Within 14 weeks, 418 Squadron had destroyed 83 V-1s, but the highest score was by the team of Bannock and Bruce who had downed 18 1/2! Once, on a single sortie, they destroyed four of the robot-bombs. Now given command of 418 Squadron, Bannock was also awarded a Distinguished Flying Cross and Bar for his efforts.

The night-fighter ace attributed much of his squadron's success to the fact that most of his pilots were former flying instructors from the BCTAP. Often, he credited his instructing experience to getting out of tight situations.[1] Dubbed 'The Saviour of London' by the British press, Russell Bannock ended the war as the RCAF's leading night-fighter pilot, with a score of 11 enemy aircraft and 19 1/2 flying bombs.

Soon sustained Allied bombing had eliminated the fixed 'buzz bomb' bases, and the Germans resorted to mobile launchers that were more difficult to detect. Obsolete Heinkel bombers had V-1s attached under their wings and were used as rudimentary 'stand-off' launchers, which also made the rockets more difficult to detect.

But even more-chilling weapons were to come. In September 1944, the first of the lethal V-2 rockets fell on Britain. There was no defence against these - the first ground-to-ground guided missiles of this century. As they were too fast for the Mosquitos, night-fighter squadrons retaliated by looking to attack their mobile launching pads.

Bannock, now commanding officer of 406 Squadron, saw his first V-2 launch in September 1944. He was patrolling over the blacked-out city of Boulogne at 2,000 feet and noticed a few trucks' headlights below. Suddenly he saw 'a giant fire-cracker come up spiral up ... into the heavens and keep going.'[2] He took a fix on the position and radioed for the rocket-firing Typhoons to destroy the launcher.

After the war, Bannock became director of RCAF Overseas Operations in London, and later attended RAF Staff College as a student. He retired from the RCAF in 1946 to join de Havilland Aircraft in Toronto as test pilot and operations manager. In 1947, de Havilland had just developed the first in its long line of short take-off and landing (STOL) aircraft, the Beaver.

On August 15, 1947, Bannock took the Beaver prototype up and put it through its paces. It combined agility with strength, power with utility — and it was proudly Canadian in conception and construction. The Beaver got its first boost in sales when the Ontario Provincial Air Service ordered 27 of them. A successful sales campaign by Bannock and his team at United States Air Force bases in Alaska ultimately resulted in the sale of 978 Beavers to the United States Army and Air Force.

The Beaver's successor, the de Havilland Otter followed in 1951, and it too was ordered by the United States Army. Bannock was promoted to vice-president of sales, and he entered his company's other STOL aircraft on the international market as they were designed. By the 1960s, the Otter, Twin Otter, Caribou, and Buffalo were operating with air forces and airlines around the world.

Russell Bannock left de Havilland in 1968 to form his own company 'Bannock Aerospace' but returned to de Havilland in 1975, becoming its president and chief executive officer. Finally in 1978, he returned to his own company.

He always remembered his night-fighter role and he held the position of president of the Canadian Fighter Pilot's Association for many years. If history is the legacy of heroes, the example of Russell Bannock will not be soon forgotten.

---

1 Barris, Ted, *Behind The Glory* (Toronto: Macmillan,1992), p. 277
2 From a speech that Russell Bannock gave at the Annual CAHS banquet on May 30, 1981.

# BARKER, WILLIAM GEORGE, vc

William Barker was first and foremost, a pugnacious survivor. Still alive after the brutality of the Flanders trenches in 1915, he not only lived through the deadly 'Fokker Scourge', flying an outdated BE2d, but managed to shoot a single-seater German fighter down from it as well. Unfortunately, because most of his victories were earned on the Italian Front, he has been overshadowed by other Canadian air aces.

Born at Dauphin, Manitoba, on November 3, 1894, Barker joined the Canadian Mounted Rifles at the outbreak of the First World War. Like so many other cavalry regiments, the men of the CMR were thrown into the

W. G. Barker, beside his Sopwith Camel in Italy. (NAM 3522)

Second Battle of Ypres in September 1915 as untrained infantry. Barker knew that his chances of living through the devastating machine-gun attacks were small, and successfully transferred to the Royal Flying Corps as a mechanic. He was posted to No. 9 Squadron at Allonville on the Western Front and he took to the air as a Lewis gunner in an obsolete BE2d. This was the Bleriot Experimental 2, a front-line reconnaissance aircraft in 1914, but now increasingly outclassed in a world of Fokker triplanes and Albatross D111s. The BE2 squadrons were slow moving targets and suffered disastrously in 1915 when the air war began in earnest.

It was Barker's skill with a machine gun in the back seat that brought a Fokker down over the Battle of Neuve-Chapelle in March 17. Promoted to second lieutenant, he was sent to England for pilot training. He proved to be an exceptional pilot, soloing after only 55 minutes of dual instruction, and getting his wings in early January 1917.

He returned to France a captain but as the new pilot he was given another old fighter to fly. The RE8 was known to be an unforgiving beast and by 1916 was outgunned by the latest generation of German fighter aircraft. To pilot one with little combat experience was suicidal. Against these odds, Barker managed to shoot down a Fokker, although he himself was almost killed when the RE8 nose-dived into a crash landing. To make matters worse, he was recalled to England to become a flight instructor.

Angry at this, Barker transferred to a scout squadron — No. 28, equipped with Sopwith Camels. After the slow BE2ds and RE8s, the Camel must have been an unnerving experience for the Canadian. It was tricky to fly. It had a huge, high-powered rotary engine that made its centre of gravity too far forward for an inexperienced pilot to control. Barker's commanding officer would write that the new pilot, used to the slow, unwieldy RE8s was very clumsy on the Camel, but he did make up for this with his shooting skills.

Barker worked hard at taming the Camel with its heavy rotary engine and short fuselage, and soon was a confirmed exponent. He even piloted a Caproni bomber behind enemy lines to drop an Italian agent by parachute. Finally, decorated by both the British and Italians, he was given command of his own squadron.

He was now 24 years old and had a reputation as a lone wolf and a gambler of his own and fellow pilot's lives. But Barker was neither. His men recalled that he shepherded the new pilots into battle and shared his triumphs with all on his flight. The Canadian loved the head-on attack, but was a caring formation leader.

On October 17, 1917 he returned to the Western Front, flying a Sopwith Snipe — the lineal descendent of his beloved Camel. He now had two aerial victories credited to him for every year of his life. For ten days with 201 Squadron over La Targette, he saw no enemy aircraft, and on October 27, took off to return to England and the boredom of a safe, desk job.

Climbing to 21,000 feet for a final look at the Front, he spotted a Rumpler reconnaissance aircraft. Barker swooped on it and sent it down in flames. That was when he ran into the full strength of Jagdgeschwader 3 and its 60 Fokkers. There was no escape and it seemed that the survivor's luck had finally run out. The incredible air battle that followed was witnessed by thousands of British and Canadian soldiers below. As the lone Snipe brought down Fokker after Fokker, the resulting roar from the war-weary soldiers in the trenches echoed across the front lines. If he could have heard them and remembered his own early experiences at Ypres, Barker would have been pleased. After a few minutes of this confusion, the Snipe had taken over 300 hits in its fuselage and Barker had lost the use of his right leg from a bullet. Fainting from the loss of blood, he went into a spin, still followed by the pursuing Fokkers. His cockpit now awash with blood, keeping control of his aircraft with his left arm, and barely conscious, Barker shot down three more aircraft before diving 6,000 feet to the ground. He crashed the Snipe into the ravaged battlefield, its undercarriage and wings torn away by the dive.

Fortunately, the aircraft did not explode, and the troops who had witnessed the uneven battle ran to pull the half dead pilot out in time. For 10 days, he lay unconscious in hospital, When he woke up, he was told that he had been awarded the Victoria Cross.

The war drew to a close as he recovered from his wounds. In civilian life, he ran an airline with fellow ace Billy Bishop, flying wealthy sportsmen from Toronto to the Muskokas. When that failed, he went into other commercial ventures until, in the 1920s, he became the representative for an aircraft manufacturer. On a snowy March 12 in 1930, Barker was testing a new aircraft, and he took off from the RCAF airfield at Rockcliffe, outside Ottawa. The plane stalled and crashed onto the frozen Ottawa River. Known to the Italians as "the artist with a pair of Vickers" Barker died instantly. The engine and fuselage of the Sopwith Snipe in which he won his Victoria Cross is now part of the permanent collection of the Canadian War Museum in Ottawa.

A Lancaster of 419 squadron, low over a V-1 flying bomb site. (NAM 24552)

# BAZALGETTE, IAN WILLOUGHBY, vc

From September 1939 to May 1945, the Royal Air Force lost 47,268 aircrew members in action. Losses were especially high in the 'round-the-clock' offensives of 1943-44. A bomber crew's first tour was 30 missions, the second 20. If the losses rose above three percent, mathematically there was no chance of survival. It was as simple as that. Yet such was the comradeship and prestige of belonging to a bomber squadron - and especially the elite Pathfinder Force — that young men willingly volunteered for tour after tour, tempting fate with every flight. Of the three men to be awarded the Victoria Cross in the Pathfinders — all posthumously — one was a Canadian, Ian Willoughby Bazalgette.

Born in Calgary, Alberta, on October 19, 1918, Ian Bazalgette attended Balmy Beach School in Toronto as a child. In 1927, his well-to-do English parents returned to England, and Ian continued his education at Rokeby, Wimbledon. At the start of the war, he enlisted first with the Royal Artillery and later transferred to the Royal Air Force Volunteer Reserve. By September 1942, Bazalgette was a fully qualified pilot officer and posted to 115 Squadron at Mildenhall, Suffolk.

The squadron was then operating Wellington bombers, and he took part in mine-laying sorties and daylight raids on enemy cities. Later that year the obsolete Wellington was replaced with the latest four-engined Lancaster. Bazalgette was awarded the Distinguished Flying Cross for a raid on Berlin, and by June 1943 was promoted to squadron leader. He completed his 30 missions and was posted out to an Operational Training Unit at Lossiemouth, Scotland. Unlike most aircrew, who would have been relieved to have survived their first tour, Bazalgette sought further action.

He recalled the day when Hamish Mahaddie of the Pathfinder Force lectured at 115 Squadron on the Pathfinder techniques. The Pathfinders were a specially formed force of experienced crews who flew their Mosquitos and Lancasters ahead of the main bomber stream to locate and mark the targets. Because they flew without the protection of the massed main formations, and because the Germans soon realised the value of shooting down a Pathfinder, their losses were horrific. Bazalgette wrote to Mahaddie and applied for a transfer.

After the first attempts were turned down, his application was accepted, and Bazalgette joined 635 Squadron, a Pathfinder unit then at Downham Market, in late April 1944. A week later he flew his first sortie to bomb the marshalling yards at Mantes Gassicourt. More attacks followed, and although no one in the squadron knew it, this was Bomber Command's prelude to the D-Day invasion.

On June 6, the invasion day itself, the Canadian helped put the shore batteries at Longues out of action, and two days later dropped eighteen 500-pound bombs on the Luftwaffe runway at Rennes. His precision and accuracy had brought him to the attention of the High Command, and for the next raid he was selected as the deputy master bomber. This meant that his Lancaster was to be the first marker aircraft over the target. In the event of the master bomber's aircraft being shot down, he would take over the complete direction of the main bomber force.

On June 12, 1944, Hitler launched his V-1 bombs in swarms against London. Too fast and too many for conventional defence, the only way they could be effectively stopped was by the bombing of their launching sites. In June, Bazalgette was made a formation marker on sorties against the V-1 sites. By now he was completely proficient at illuminating the target areas and on the July 20 was chosen to be the master bomber for daylight attacks in support of the Allied beach-head invasion forces.

On Friday, August 4, the Pathfinder Force mounted another daylight raid on a V-1 storage depot at Trossy St. Maximin. The site had already received the attentions of RAF bombers during the previous two days, and the German flak crews had been reinforced. When, as a prelude to the raid, the Pathfinder Mosquitos screamed across the site, the anti-aircraft guns immediately put up a thick curtain of flak.

While Bazalgette was not the master bomber on this occasion, he was in the spearhead of the 14 aircraft from his squadron. His crew were all operational veterans, three of whom had been with him since the OTU days at Lossiemouth.

When the aircraft of both the master bomber and his deputy took direct hits, Bazalgette assumed command. His Lancaster began the run-in over the target, now a wall of murderous flak. Ignoring the shells that his aircraft took, he concentrated on the markers left by the Mosquitos. Within seconds, the Lancaster was grievously wounded; both starboard engines were aflame, as were the wings and part of the fuselage.

The bomb aimer had his arm and part of his shoulder torn away, and the mid-upper gunner was overcome by fumes from the raging fire within the aircraft. Doug Cameron, the rear gunner, was able to swing his turret to beam and saw that the fire on the starboard wing had reached the aircraft framework. Worse, he noticed a wave of petrol slopping along the fuselage toward him.

Bazalgette ignored the flak and continued to concentrate on dropping his markers over the target for the others. Only after he had done so did he attempt to take the burning Lancaster out. But the bomber, crippled without engines or a starboard wing, was losing height rapidly and the order was given to bale out.

Knowing that they could not take the mortally wounded bomb aimer or suffocating gunner with them, the other crew members were reluctant to jump, but they obeyed their skipper. With the stricken Lancaster still 1,000 feet up, Bazalgette could have parachuted to safety at that point, but he chose to attempt a landing and bring in the two remaining aircrew on board. Even this might have been successful, except that on final approach, he saw the little village of Senantes directly in his path.

The villagers watched as the burning Lancaster plunged towards them and, at the last possible second, turn away from their homes to execute a perfect landing in a nearby field. Then it exploded.

The three crew members who had jumped were hidden by the French Resistance and eventually liberated by the advancing Allies. The grateful inhabitants of Senantes found Bazalgette's remains in the wreckage. After liberation, they paid him the simple but sincere tribute of burying the pilot among their own families in the local churchyard. By sheer coincidence, his sister Ethel in American Intelligence, was posted nearby and was able to attend her brother's burial. On December 18, at an investiture at Buckingham Palace, Ian Willoughby Bazalgette's mother received her son's Victoria Cross.

At the inquest, the most fitting tribute to their skipper came from two members of the crew. The rear gunner, Doug Cameron, described Bazalgette as "a magnificent pilot and a born leader, adored by all his crew," while Geoff Goddard, the navigator, said "... an excellent pilot, cool, efficient, with the capacity to inspire confidence in all who flew with him ... popular with both ground and air crews."

Matt Berry will always be associated with the Junkers. The original 'flying box car' sometimes flew with canoes and floats lashed beneath its solid wings. (NAM 8096)

# BERRY, ARTHUR MASSEY (MATT)

Bush flying in the early decades of this century was a hazardous business for little material reward. Unlike the race to conquer the Atlantic, there was no glory in pioneering aviation routes over the Canadian north, or much potential for commerce. To survive in the business required a special type of person, one who was resourceful, persevering, and courageous. Typical of this breed was Matt Berry, enshrined in the annals of bush flying as 'King of the Northern Fliers.'

Arthur Massey Berry was born on June 19, 1888 in March township near Ottawa, Ontario. When the First World War began, he enlisted in the 153rd Battalion of the Canadian Expeditionary Force but once overseas, in 1917, transferred to the Royal Flying Corps. He became a pilot and was posted back to Canada as a flying instructor with the RFC, to be based at Deseronto, Ontario.

Berry returned to farming after the war and bought a large farm in Alberta. But falling crop prices soon bankrupted him and he was forced to sell out. Recalling his RFC days, he thought about becoming a pilot and took his commercial licence in 1928. 'Doc' Oaks, the director of the Northern Aerial Minerals Exploration (NAME) company valued the training that the RFC gave and hired him. At a time when the public saw an aircraft as little more than a barnstormer's toy, NAME was in the throes of demonstrating its value in mineral exploration and rural mail delivery.

Using a Fokker Super Universal initially on flights from Sioux Lookout, Ontario, and later from bases in the provinces of Alberta and Manitoba, Berry flew as far afield as British Columbia. In his first summer, he and T.M. 'Pat' Reid carried prospectors with their supplies to northern Hudson Bay. Leaving Winnipeg, he stopped at The Pas, Fort Churchill, Cape Eskimo, Chesterfield Inlet, and Baker Lake. There he remained to ferry the prospectors about before the winter freeze began. Even at the height of summer, there were the hazards of devastating storms, little fuel, and the precaution of having to beach the Fokker every night. In spite of this, the only accident occurred at Churchill on the return trip, just before the river began to freeze. Taxiing to take off in a heavy sea strewn with ice floes, Berry damaged a float and the Fokker had to be beached for repair. It had been an eventful summer.

In the early 1930s, Berry made several trips to the Coppermine River area and Great Bear Lake, hauling miners and their equipment back and forth. Yet despite the many contracts, it was not a profitable enterprise and NAME sold out to Canadian Airways in 1932. Berry took the opportunity to join the newly formed Mackenzie Air Service.

With them, in the winter of 1933-34, he flew to Gjoa Haven in King William Land to bring back a cargo of furs for the Hudson Bay Company. The historic flight was made in conditions that would have daunted a lesser pilot. The temperature was never above -45 degrees F., and the magnetic compass was all but useless. The sea was filled with old ice from the northern channel, making all landings for his Fokker Universal treacherous.

A year later, while landing near Cambridge Bay, his aircraft hit a boulder under the snow and lost a ski, stranding Berry and his passenger, the inspector of all the Hudson Bay Posts. When this was announced on the radio, aircraft flocked to their rescue, causing Berry to wonder if it was because of him or his passenger's importance.

In 1934 while taking part in the Edmonton and Northern Alberta Aero Club's annual Air Day, Berry was seriously injured when the faithful Fokker went out of control and crashed, killing his engineer. An examination revealed that during a recent overhaul, some of the control cables had been incorrectly refitted.

After a convalescence, he joined Canadian Airways to work for 'Punch' Dickins. He was reluctant to fly again, thinking that after the crash, no one would ever be his crew or passenger. But Dickins convinced him to return and sent him off to take a refresher course at RCAF Camp Borden, which emphasized instrument flying. Immediately after that, Berry went to the RCAF base at Rockcliffe, Ontario, to learn about the latest in navigational aids — radio beacons.

On his return to flying, the bush pilot was given the aircraft he is most associated with — the Junkers CF-ARI. This was the rugged 'Flying Boxcar' that could be flown anywhere, loaded to the door, even lashed with canoes and floats under the wings. It was a forgiving workhorse that Berry used for the many rescue missions he is remembered for.

Isolated communities of trappers, miners, and missionaries depended on being supplied by air. The Junkers and Bellancas of the bush pilots were their grocery vans, newspaper boys and hardware stores in one. Bush pilots hauled every possible item in their cargo holds — from darning needles to generators, fresh meat to dynamite. There were communities in dire need of fresh milk, and many a dairy cow was coaxed into the hold and lashed down for a somewhat noisy flight. Others were desperate for news — whether in newspapers or letters — and met the pilot, clamouring for both. One bush pilot liked to tell of the time when he dropped a load of letters tied in a bundle over a small village. On the way down the strings broke and the letters fluttered through the forest like flocks of ptarmigan. He later heard that the whole village spent days hunting for them. Another remembered dropping what he thought was frozen meat to a remote camp at the onset of winter. He later learned that it was a collection of records for the camp's victrola. Only two of the three hundred records survived and they were played all winter long.

One of the most spectacular missions Berry was involved in occurred in August 1936 with the RCAF. Flight Lieutenant S. Coleman and Aircraftsman Joseph Foley had just delivered engine parts to another RCAF crew stranded at Hunger Lake, NWT, and were returning to Fort Reliance when a sudden storm threw them off course. They landed at a small lake and left a note in an empty oil drum. In it they explained that they would fly in a southernly direction until their fuel gave out. They then did so, finding another lake when their fuel ran out and waited for rescue.

The RCAF dispatched six aircraft to search for Coleman and Foley, but as the days gave way to weeks and they were not found, civil pilots like Matt Berry were called in. Over 70,000 miles were covered by the time the summer had ended. Then on September 13, with the help of some trappers, the oil drum with the note was discovered. At the same time, winter had started in with freezing temperatures and storms blowing down from the Arctic. As all the aircraft were on floats and not skis, they were withdrawn to be refitted, causing further delay.[1]

Having been in similar situations, Matt Berry followed what he called 'a hunch'— an instinctive guess about what he would have done given the amount of fuel that the RCAF fliers may have had. On September 16, 1936, he located Coleman and Foley. Without rifles for game, the airforce fliers had survived on boiled squirrel and berries. Although rescuers expected them to be found due south and had concentrated the air search in that area, Berry realized that their compass would have been malfunctioning and flew instead in a northwesterly direction. It was a gamble, but for this outstanding rescue, he was recognized with an award from the City of Edmonton.

That year, Berry was also called in to fly out the stranded passengers of the schooner *Our Lady of Lourdes* off the Arctic coast. These were missionaries with Eskimo children and their dogs who had left the ship and started out toward the nearest harbour in the hope of rescue. Berry departed from Aklavik in his ski-equipped Junkers for where he thought the harbour would be.

Because of the lack of daylight during the Arctic winter, navigating by landmarks is almost impossible and all flying is done by instruments. There was a constant gale and the snow on the ground was whipped up to a height of about 1,000 feet obscuring the view. For all of the flight, visibility was less than 25 yards ahead of the Junker's propeller, and Berry had to keep using his flashlight to check his instruments. Against all odds, he found the missionaries and landed. The priests and the children were in a pitiful state as marauding polar bears had cleaned out their food cache and the survivors were going to eat the dogs soon. The entire party including the dogs, were bundled into the Junkers and flown off during a break in the weather. But as it was mid-December, even Berry and the Junkers could not cope with the sudden blizzard that sprang up on the way home, and they were forced down on a lake near Aklavik to wait it out for a number of days.

For these and other rescue flights, 'Matt' Berry was awarded the Trans-Canada Trophy in 1937. Fittingly, the award was made at the Canadian Institute of Mining in Montreal with Air Vice Marshal W.A. 'Billy' Bishop, VC. presiding.

When the Second World War broke out, Berry immediately volunteered but, because of his age, was turned down. He was working at an Air Observer School at Portage la Prairie, Manitoba when Grant McConachie of Canadian Pacific Airlines recruited him in 1942 for the CANOL Project.

CANOL or Canadian Oil Line was a joint defence undertaking between Canada and the United States to carry oil from the Norman Wells oilfields to Whitehorse for wartime purposes. With its entry into the war, the United States was looking for additional fuel supplies for its expanded industries. The construction of a network of pipelines meant that communities and airfields would have to be built at strategic points along the Athabasca and Mackenzie rivers. McConachie's Canadian Pacific Airlines was awarded the contract to fly men and supplies in to build the landing strips.

Appointed superintendent of construction, Berry worked through the remainder of the war bringing in personnel and equipment to Fort McMurray, Embarras, Fort Smith, Fort Resolution, Hay River, Fort Providence, Fort Simpson, Wrigley, Canol Camp, and Norman Wells. He became the prime contractor for airport construction after the war ended, buying the surplus U.S. equipment left behind.

In 1947 Berry became president of the Davenport Mining Company of

Toronto, Ontario, and in 1949 operated an air charter service through the Northwest Territories. He sold his flying operations in 1951 to devote himself completely to mining exploration. After a lifetime of service to flying in the North, he died in Edmonton, Alberta, on May 12, 1970.

1 Ellis, Frank, *Canada's Flying Heritage* (Toronto: University of Toronto Press, 1954), p. 337-338.

Sent back to sell war bonds, 'Buzz' Beurling was uncomfortable as a hero and longed to return to the action. (NAM 14234)

# BEURLING, GEORGE FREDERICK (BUZZ)

Other RCAF pilots refused to fly with 'Screwball' Beurling, disliking his callous approach to aerial combat. Yet, disdained by his own country's air force, Beurling, happiest in the air, became the greatest Canadian fighter ace of the Second World War.

'Buzz' Beurling was born in 1921 in the Montreal suburb of Verdun. His love of flying began at six years of age when his unsuspecting father built him a model aircraft. At 10 he read every book and every comic on flying that he could get his hands on. He knew by heart the exploits of the First World War fighter pilots, and spent every spare minute watching the aircraft at the local airfield. George's parents wanted him to be a commercial artist like his father, or better still, study medicine at McGill University. The depths of the Depression were no time to think of a career in aviation.

Just before his 11th birthday, on his way to the local aerodrome Beurling was caught in a violent rainstorm and sheltered in a hanger. A kindly pilot feeling sorry for the wet, shivering boy and knowing of his enthusiasm, offered to take him on a flight if he got his parent's consent. Mr. and Mrs. Beurling thought that this was a joke and agreed.

After that memorable flight, every penny, every breath George had was devoted to taking flying lessons. He sold newspapers on street corners and built model aircraft to sell to classmates — anything to pay for flying lessons. At the age of fifteen, against his parent's wishes, he quit school and got a job to increase his income.

A year later, he had 150 flying hours in his log book and passed all the examinations for a commercial pilots licence — only to be told that he was too young to be licensed. Desperate to acquire flying experience, he set out for China — hoping to join the Nationalist Chinese Air Force then reeling after the Japanese onslaught. He crossed the U.S. border, thinking to head for San

Francisco and then work a passage over to China and sign up. Unfortunately, he was arrested as an illegal immigrant at the border and sent back home.

By then it was September 1939 and the Second World War had broken out. Beurling immediately applied to the Royal Canadian Air Force as a pilot — only to discover that he lacked the necessary school-leaving certificate. He then volunteered for the Finnish Air Force, which was recruiting experienced pilots for their bitter Winter War against the Soviet Union. He was accepted, but he still needed his father's permission to leave the country. Mr. Beurling had not forgiven his son for quitting school and the permission was refused. George now at his wit's end, continued to do what he loved best — fly privately. By April 1940, he had over 250 hours of solo flying recorded in his log book.

After Dunkirk, the young Montrealer, like the rest of the world, knew that it would be Britain's turn next and that the Royal Air Force would need experienced fighter pilots. He signed on as a deckhand on a Swedish ship bound for Glasgow and once there reported to the nearest RAF recruiting station. Sadly, he was told that to be considered for the RAF he would need his birth certificate and his parents' consent, neither of which the youth had. He immediately sailed back across the U-boat infested Atlantic, obtained both and crossed over once more.

Finally accepted, he was sent to the RAF basic training school at Buxton-on-Sea, where the ground instructors, aware of this 'colonial's' lack of education, made the classroom a hell for him. But on September 7, 1940, he was selected for pilot's training, and a year later reported to 403 Squadron with his wings. Incredibly, even at this late stage, Fate seemed to thwart Beurling. Because he had been trained by the Royal Air Force, when 403 Squadron was made an all-Canadian fighter squadron, he was posted out of it. Beurling chaffed at the King's Regulations and at the lack of combat opportunities in his next squadron and applied for an overseas posting.

That summer of 1942, the combined might of the Luftwaffe and the Regia Aeronautica were pounding the British Mediterranean colony of Malta. With its Sicilian bomber airfields only 70 miles away, the Axis expected that the besieged island would soon have to surrender. Surrounded by unfriendly territory, the relief of Malta required ingenious planning.

Carried away by the audacity of it, Winston Churchill himself endorsed the imaginative idea of reinforcing Malta by positioning aircraft carriers outside the Luftwaffe's bomber range. The plan called for Spitfires to be flown off the carriers to the island. On June 9, 1942, Beurling took off from the flight deck of the aircraft carrier HMS *Eagle* in a factory-fresh Spitfire Mk.V equipped with drop tanks. As soon as they landed, the new pilots were greeted by a Luftwaffe air raid and Beurling watched from his slit trench with excitement as the Ju88s pounded the airfield. At last, he was in his element!

That afternoon, 249 Squadron was scrambled to intercept a bombing raid over Gozo, the neighbouring Maltese island. Beurling's section went after 20 Ju88 bombers. Then from 18,000 feet, the Messerschmitt pilots, who had been lying in wait, jumped the newcomers. Beurling excitedly sprayed burst after burst of machine-gun fire at the weaving fighters. In the ensuing melee there was no time to confirm a hit. The object was to prevent the bombers from getting to Valetta harbour and sinking a convoy that had limped in.

The Canadian's first kill was an Italian Macchi fighter; next, a bomber fell to his guns. Later, patrolling over the RAF airstrip of Safi he heard a desperate call on his R/T for aid. Beurling later claimed it was a lucky shot but his cannon fire brought down his first Messerschmitt, then he landed for fuel. An hour later, 249 was scrambled again as over 30 Junker Stuka 87s were attacking the ammunition ship *Welshman* in the harbour. Beurling joined in the fray shooting down an escorting German fighter and a Stuka, then he crash-landed near the cliffs. Thus ended his first day on Malta.

Through that desperate Mediterranean summer, Beurling's score began to mount as Messerschmitts and Macchis fell to his guns. His success was attributed to skills common to what had made Canadian pilots famous during the First World War: marksmanship and a disregard for regulations. Like Bishop and Barker in 1917, Beurling was happiest as a lone wolf, an average pilot who made up for his mundane flying skill with aggression and crack shooting.

Unlike the affable Bishop, when not fighting, Beurling was morose and eccentric, and his RAF colleagues nicknamed him 'Screwball' for his strange ways. The story went that he would throw a piece of meat on the ground and then stamp on the flies that settled on it, muttering "Goddam screwballs!"[1] He carried a Bible with him in combat, yet murdered his opponents as they parachuted out of their aircraft.

Worse still, in the eyes of the RCAF, he twice refused a commission. Always insecure about his lack of a formal education, he was in his own mind, not the officer type. The press both at home and in London, lionised him as 'The Defender of Malta.' Finally, the High Command overrode 'Buzz's' protests and made him an officer.

By September 1942, with Malta-based torpedo bombers sinking Rommel's supply ships regularly, the Luftwaffe made a final determined effort to destroy the island's defences. So fierce was the subsequent air fighting that on the single day of October 9 Beurling shot down three Messerschmitts and one Junkers 88. By the end of his second tour on October 13, his score stood at 24.

Shot down and wounded, he was sent to England to recover. En route home, as his transport plane neared Gibraltar, Beurling sensed impending disaster. He moved to one of the emergency exits just as the big Liberator stalled fifty feet out over the sea. The fighter pilot jettisoned the emergency hatch and dived out. The Liberator crashed into the water, and Beurling one of the few survivors, swam one hundred and fifty yards to shore.

The pilot who had been refused entry into the Royal Canadian Air Force, now returned to Canada a national hero. In 1943 he went to England and was ordered to attend an investiture at Buckingham Palace to receive from King George VI, the DSO, DFC, DFM and the Bar to the DFM — all at the same time.

Now a recognized air ace, Beurling was assigned to 'Johnnie' Johnson's 412 Squadron. But the 'big wing' concept of 36 Spitfires sweeping over France did not appeal to the lonely hunter. Johnson recognized Beurling's talent for deflection shooting and wanted him to pass this skill onto the new pilots. After the hectic Malta battles, the fighter pilot hated the gigantic 'big wing' sorties that invariably failed to encounter a single Messerschmitt. The thoroughbred Spitfires were relegated to ground attack duties as the Luftwaffe had disappeared from French skies. All the pilots disliked these as a single stray bullet in the Spitfire's glycol tank would cause the engine to seize up.

Yet the Luftwaffe had some teeth left — the sharpest one being the lethal Focke Wulf 190 fighter. But before the High Command sent him to the gunnery school at Catfoss, Beurling managed to shoot down four of the new German fighters. Always self-sufficient, he found it difficult to teach his brand of marksmanship to pilots trained on orthodox methods.

By war's end 'Buzz' Beurling was a squadron leader and officially transferred to an embarrassed RCAF. After being demobilized, he drifted from commercial flying to stunt flying for a living. He even became an insurance salesman. Never a civilian at heart, 'The Defender of Malta' had only been truly happy in an aerial dogfight.

Then, in 1948, he heard that the emerging state of Israel was being invaded on all sides by its Arab neighbours. The Israelis were scouring the world for Second World War fighter aircraft and pilots. As he had done for the Chinese and Finns, Beurling quickly volunteered. Several ex-RCAF pilots did the same and to his delight Beurling heard that the Israelis were flying Spitfires.

Two days after the Israelis declared independence, Beurling was already on his way to Tel Aviv. On May 20, he was ordered to fly a Canadian-built Noorduyn Norseman full of medical supplies from the Rome airport to the new state. Beurling was unfamiliar with what was a slow bushpilot's aircraft, and a former Canadian naval pilot had offered to check him out on it. The

Norseman took off, circled the field, and approached to land. Those on the ground saw it overshoot the runway and climb again. A few seconds later, it dived into the ground, killing both pilots instantly. At the age of 26, 'Buzz' Beurling had flown once too often and could cheat Death no longer.

Never the public's idea of an air ace, wholly uncomfortable outside a Spitfire cockpit, 'Buzz' Beurling had outlived his time.

---

1 Jackson, Robert, *Fighter Pilots of World War II* (New York: St. Martin's Press, 1976), p. 66.

# BISHOP, W.A. (BILLY), VC

O f all the Canadian fighter aces of the First World War — Collishaw, MacLaren, Barker, and Claxton, it was Billy Bishop that achieved immortality. Born February 8, 1884, at Owen Sound, Ontario, William Avery Bishop was an accomplished rider and marksman long before being accepted at the Royal Military College, Kingston, in 1911. Remembered as a somewhat ill-disciplined cadet, by the outbreak of war, Bishop was commissioned as a lieutenant. He enlisted in the Mississauga Horse, the cavalry detachment of the Second Canadian Division.

W.A. 'Billy' Bishop in his Nieuport 17. (NAM 11555)

W. A. Bishop (left) and W. G. Barker (right) with war booty. In 1919, they would start an air charter company, flying wealthy sportsmen from Toronto to the Muskokas. (NAM 8069)

Perhaps his disillusionment with the cavalry began on the Atlantic crossing, when he sailed on a cattle boat with 700 seasick horses. As bad as the *Caledonia* was (it almost went to a watery grave off the coast of Ireland), as within full sight of all on board, a U-boat surfaced and calmy proceeded to sink three ships in the convoy, before the destroyer escorts could intervene.

Years later, Bishop would write that he resolved to transfer to the Royal Flying Corps while on manoeuvres on Salisbury Plain in England. As the military came to grips with this new form of warfare, he seemed to spend more time shivering in muddy trenches than on dashing cavalry charges. It was on an occasion such as this that he saw far above the mud and manure, an aircraft fly over in, what seemed to him, a god-like, detached manner. It was at that precise moment he resolved to apply to the Royal Flying Corps.

The only way of entry into the RFC was as an observer, and in the autumn of 1915 he went to France as one with 21 Squadron. For four uneventful months he never fired a shot, and he even ignobly suffered a crash landing. In it, Bishop injured his knee and was sent to a hospital in England.

Upon discharge, he ecstatically learned that he was to be trained as a pilot. His first solo was a disaster and he almost destroyed the old Farman trainer on landing, a manoeuvre he always had trouble with. However, his riding ability seemed to make him a natural pilot and he earned his wings after only 15 hours of solo flying. Then to his disappointment, instead of being posted to the Front, Bishop was relegated to a Home Defence Unit to fly BE2cs against the Zeppelin menace.

Finally on March 17, 1917, Bishop joined 60 Squadron at Izel-le-Hameau in France. There he flew Nieuport Scouts, nimble French-built fighters which by 1917 were adequately armed, with a Lewis gun mounted on the overwing rail and a synchronized Vickers .303 firing through the propeller blades.

Unfortunately, his haste to fly combined with a somewhat hamfisted approach dogged Bishop even at this late stage, for he still had extensive problems with landings. There were burst tires, strained airframes and on his second Nieuport flight, in a particularly clumsy attempt, he wrecked the aircraft beyond repair. But Bishop (and his squadron commander) persisted, and he remained with the squadron.

Then on March 25 he was inexplicably made a leader of a five-man offensive patrol. At 9,000 feet his flight was attacked by three German Albatross D11s. Bishop dived after one, firing a tracer into it. It fell into a spin and he followed it down. The Albatross crashed, and Bishop pulled his aircraft out of his dive only to find that as the plugs had become oiled his radial engine had died and he could not restart it. Luck was with him and he was just able to glide back over the Front, into the Allied lines. There he cleaned the plugs with a toothbrush and took off again. Five days later, he shot down another Albatross and the following week three more. Now the 'squadron ace', Bishop was given the privilege of taking off on lone, roving patrols over the Front.

That month of April saw the start of the battle of Arras. The Allies' offensive began with a great artillery barrage all along the Front. To prevent the enemy from locating the British advances, 60 Squadron was ordered to shoot down as many German observation balloons as possible. Nicknamed 'sausages' by the RFC pilots, each balloon was the eyes of the German Army, and heavily protected, not only by patrolling fighters but also by murderous flak batteries that threw up what were called 'flaming onions'. These flak guns were quick-firing, former motor-torpedo boat armament that shot tracer balls up to 9,000 feet into the air. Bishop not only shot down one balloon's defending fighters but, ignoring the flak, the balloon as well.

For this achievement he was decorated with the Military Cross. His score now began to rise. During that April and May he shot down 20 aircraft and was promoted to captain. It was, however, on June 2 that he performed the single act that would forever ensure his place in the annals of air warfare. He had always valued the tactics of surprise and now he planned the ultimate one. It was to be a dawn raid on a German airfield — what later military strategists would describe as 'a pre-emptive strike.' The plan had actually originated from a suggestion made by Albert Ball, a visiting British ace who advised that as a low-level attack on a German airfield had never before been attempted, the lone attacker might just get away with it.

The element of aggression appealed to Bishop and he chose an enemy airfield southeast of Cambrai. With the tactics of surprise in his favour, he hoped to rake the sleeping German aerodrome from end to end and escape home before their fighters could catch up with him. His combat record tells what happened in the few minutes he was over the chosen enemy airfield:

> I fired on 7 machines ... some of which had their engines running. One of them took off and I fired 15 rounds at him from close range 60 feet up and he crashed. A second one taking off, I opened fire and fired 30 rounds at 150 yards range, he crashed into a tree. Two more were then taking off together. I climbed and engaged one at 1,000 ft. finishing my drum and he crashed 300 yards from the aerodrome. I changed drums and climbed East, a fourth H.A. came after me and I fired a whole drum into him. He flew away and I then flew 1,000 ft. under 4 scouts at 5,000 ft. for one mile and turned West climbing.[1]

He shot down four of the enemy before the groundfire became increasingly accurate, and then made his escape. The furious Germans chased the bullet-ridden Nieuport now with a very nauseous Bishop in it, back across Allied lines. After some dangerously low flying, he managed to evade them and land at his own airfield, the Nieuport shot full of holes.

For this act of supreme heroism he was awarded the Victoria Cross on August 11, 1917. This was also the only case of such a high honour being awarded on the word of the recipient alone, but so well known was Bishop's record for surprise and accuracy that his commanding officer believed him implicitly.

Only a fair pilot, Billy Bishop made up for this by being an expert deflection shot. This was a technique where the pilot judged the speed and heading his target was flying and fired at a point directly in front of it. If done correctly, both the enemy aircraft and the stream of bullets arrived similtaneously.

His marksmanship stemmed from his boyhood hunting days in the Ontario woods. Because the Nieuport was slower than the German fighters, deflection shooting proved an asset. Coupled with this was an absolute hunger for recognition.

Bishop's drive for fame combined with his shooting skills soon drove his score to 45 kills. On April 30, in a space of two hours before lunch, he reported 8 combats against 19 enemy aircraft. Now promoted to major, he was awarded a bar to his DSO and sent to London where he received the Victoria Cross from King George V. Then he sailed from Liverpool for Canada.

At home he was feted by the public, crossing Canada on a recruiting campaign and even finding time to get married. Early in 1918, he returned to England as chief instructor at the Aerial Gunnery School to pass on his skills. But with the final German offensive at its height, on March 13, 1918, Bishop was given command of his own squadron — No. 85, and returned to the Front. High Command realized that the Canadian air ace was more valuable as a living symbol, and rather than risk his life as a fighter pilot, ordered him to a desk job in London. They grudgingly gave him a final fortnight of flying before being grounded forever. Bishop used the remaining time to launch what became known as his 'Carnival of Death.' In those two weeks, he shot down 25 of the enemy, 12 of them in the last three days. In just over a year, he had a score of 72 enemy aircraft. On June 19 he scored his last victory downing a two-seater fighter.

In a twist of fate when he transferred out of the new Royal Air Force to the Canadian military, he was made a temporary lieutenant-colonel in the Canadian Cavalry. Canada's best-known air ace went home to begin work on the creation of a wholly Canadian Flying Corps. Constantly harried by the British who did not want to see an independent air force — and by the Royal Canadian Naval Air Service — who considered themselves as the senior (by two weeks) service, the campaign for a national air force needed just such a high profile spokesman.

By 1918, twenty-five percent of all the RAF flying personnel and forty percent of all the aviators on the Western Front were Canadian. As the most famous Canadian airman of all, Bishop could take his views to the highest levels and all the newspaper-reading public. He emphasized that although his countrymen were doing well enough under British command, to put them in a Canadian Corps — and eventually work in cooperation with Canadian troops — would instill in them an unrivalled esprit de corps. It would also form the nucleus of a post-war Canadian air arm. By the time Bishop returned to France to implement his program, the war was over.

After the war, like hundreds of former fighter pilots, he tried various com-

The proudest moment in the life of a BCATP graduate was to be presented with his wings by Air Marshal 'Billy' Bishop himself. (NAM 20350)

mercial enterprises around the country. The former ace set up one of the first airlines in Canada, with fellow fighter pilot William Barker. Bishop-Barker Aeroplanes Ltd. flew wealthy Toronto sportsmen to vacation resorts in the Muskokas, but while both men had a lot of fun doing it, neither had any business acumen, and the airline was soon bankrupt.

Bishop lectured and wrote extensively about his wartime experiences but remained committed to the concept of a wholly Canadian Air Force. He became more active in this as the world moved towards another war. In 1938 Billy Bishop was made an honorary Air Vice Marshal of the Royal Canadian Air Force, a public relations role he relished. He was also named chairman of the Air Advisory Committee to the Minister of National Defence.

The day after war was declared, Bishop flew to New York, rented a hotel room and called up all his First World War contacts. Then he launched a recruiting drive aimed at tapping the vast pool of pilots in the neutral United States for the RCAF.

During the war, he served as the director of the RCAF, pinning the coveted 'fliers wings' on hundreds of young graduates of the British Commonwealth Air Training Plan. The First World War hero was now an ebulient, overweight honorary Air Marshal whose reputation to carouse was legendary.

Of the many wartime public relations activities he engaged in, Bishop is probably best remembered for his acting role in the Hollywood movie *Captains of the Clouds*. Filmed at Uplands Air Training Field near Ottawa, it starred Jimmy Cagney as the heroic bush pilot who flies in to Canada to join up. The movie was a joint RCAF/Hollywood studio effort to glamorize the war effort and recruit American fliers into the BCATP. Receiving high praise for his acting from the movie director Hal Wallis, Bishop's swan song was the stirring speech he gave at the graduation scene at Uplands to an authentic class of Air Training Plan pilots.

Billy Bishop died peacefully in West Palm Beach, Florida, on September 11, 1956. His body was flown home to Toronto and thousands of his countrymen lined the city streets as his funeral cortege passed by. Overhead, a squadron of Canucks dipped in homage. It was his Canadian air force, flying Canadian-built jets, confident and secure in the traditions that he had fought for.

Of him, fellow air ace Eddie Rickenbacker said: "Billy Bishop was a man absolutely without fear. Others avoided, if they could, combat. He went looking for the enemy. He was the bravest man I ever knew."

---

1 Wise, S.F., *Canadian Airmen and the First World War* (Toronto: University of Toronto Press, 1981), p. 414.

# COLLISHAW, RAYMOND

S econd only to Billy Bishop as Canada's greatest fighter pilot, and the top naval air ace of the First World War, Raymond Collishaw achieved fame as the leader of the most famous Allied fighter unit of the war — the Sopwith Triplane 'Black Flight.'

Born at Nanaimo, British Columbia, on November 22, 1893, Raymond Collishaw spent his life looking for adventure. While still a teenager, he sailed as second mate aboard a tramp steamer, and in 1911 served with the famous British explorer R.F. Scott in the Antarctic. At the outbreak of war, he transferred from the Fishery Protection Service to the British Royal Naval Service.

Raymond Collishaw in the cockpit of a Sopwith Triplane. (NAM 9781)

The only way a young man could be accepted as a pilot at this early stage of the war was to train at his own expense and Collishaw learned to fly at the Curtiss Flying School in Toronto.

In August 1916 he joined No.3 Wing, Royal Naval Air Service, the so-called 'Sopwith Sailors' then based at Luxeil-les-Bains in France. This was not only the first British long-distance bombing wing, but because it became operational just as the first batch of Canadians had finished their training, it also had the highest Canadian participation of any formation in the war. It specialized in one thing: strategic bombing. This means of warfare had not attained the lethal capacity to kill thousands of civilians as it would in the Second World War — but even then was unpopular with the press and public. As a result, Collishaw's victories with No.3 Wing never achieved the fame that other Canadian air aces did.

Spurred on by Winston Churchill, the RNAS dropped explosives deep into enemy territory, their primary targets being zeppelin sheds and railway yards. On October 12, 1916, flying a Sopwith Strutter on a 223-mile round trip with other aircraft, Collishaw shot down his first Fokker.

An impressive fighter for its day, a Sopwith was heavily armed but weighted down with two 112-pound bombs. Then current bomb-sighting equipment and techniques required that a bomber approach the target directly up or down wind. While the pilot did this, the bomb aimer used a stopwatch to measure his speed by two sightings of one object on the ground, then setting his moveable foresight to correspond with the time/distance interval between the two measured sights. All things being equal, the bombs were then released over the target. The complicated procedure meant that the Sopwith was easy prey, not only to ground batteries, but also roving enemy fighters.

Indeed, so new was the science of long-range flying that pilots frequently became lost in the clouds or mountains. Collishaw himself wrote that on one occasion having evaded his pursuers successfully when he was sure he was behind Allied lines, he landed at an aerodrome. It was only while taxiing among the parked aircraft that he realized they all had the German Iron Cross markings! He quickly jammed the throttle forward and took off.

He managed to shoot down two more fighters while on bombing runs before being transferred as flight commander to No.10 Naval Squadron at Furnes on the Flanders coast.

Now he was able to hand-pick four other pilots — all Canadian, all top scorers, for his 'Naval 10' Flight. They flew Sopwith Triplanes — the engine cowlings, wheel disc covers, and metal forward fuselage painted black with the aircraft names in white lettering. This was done so that the mechanics could recognize their own charges and go immediately to their assistance when they returned from patrol. Other flights were identified with red and blue colours. The infamous 'Black Flight' aircraft were named *Black Maria, Black Death, Black Sheep, Black Roger,* and *Black Prince.* All were experienced pilots, the youngest of the flight was only 19.

The Canadians quickly established themselves as the foremost fighter squadron on the Front. As the RFC's aerial offensive over Arras was being blunted by the determined German Albatross Jastas, the area around Dunkirk where the 'Naval Ten' flew became a private Canadian hunting ground. The Sopwiths were superior to the Albatross and Halberstadt scouts, and the black triplanes acquired a fearsome reputation. By June 5 Collishaw had shot down 13 aircraft — the next day he downed three German fighters in a single action. By June 15, his victory total stood at 23.

The Germans retaliated by sending Jasta 11 against the 'Black Flight'. Led by the legendary Manfred Freiherr von Richthofen, already credited with 56 victories, this was the cream of not only the German air force but a generation as well. With the German aircraft painted a gaudy scarlet circling the black Sopwiths, it must have been a battle in technicolour. One of the 'Black Flight' was shot down, but Collishaw sent von Richthofen's deputy down in flames. For three weeks the RNAS Canadians and Jasta 11 battled it out. Tactically the Germans held the edge. It was the Allies' numerical advantage that finally broke their attacks.

By July 28 Collishaw's score stood at 37, and he returned to Canada on leave. On his return to France in November, after a spell as an instructor, he joined the RNAS Seaplane Defence Station at St. Pol. In December, he was given command of 13 (Naval) Squadron equipped with the latest British fighter, the Sopwith Camel. With it, he was able to quickly shoot down three German aircraft. During the Ludendorff Offensives of 1918 when the German armies determinedly pushed towards the Channel, the Allies were able to exploit their air superiority. All squadrons were pressed into bombing, strafing, and artillery spotting to stem the advance. Collishaw, now commissioned as a major in the Royal Air Force, was placed in command of 203 Squadron. On a single day, his flights dropped 196 bombs and fired 23,000 rounds while attacking German infantry. He hated sending his flights out on ground-support work, realizing that it took a single enemy bullet to bring down a good pilot.

Through the summer, he was back in action, scoring more victories against the increasingly demoralized German air force. Too late to have a measurable effect on their declining fortunes, the Germans belatedly introduced what was perhaps the best fighter aircraft of the First World War — the Fokker DVIII. With a better rate of climb than the Camel, it was more manoeuvrable than even the new Sopwith Dolphin. The Allies thought it so good that the Armistice terms specifically ordered the surrender and destruction of all DVIIIs. In spite of this Collishaw shot down 10 Fokker DVIIIs, his experience in tactics telling against the new, half-trained German pilots.

With the enemy falling back in all areas, Collishaw was withdrawn from the Front and, like Bishop, sent home to plan the foundations for a Royal Canadian Air Force. On October 1, 1918, at the age of 25 he was promoted to lieutenant colonel. When the squadrons of the new Canadian Air Force were constituted, Billy Bishop chose Collishaw to command 1 Squadron, but he was unavailable.

As if being a fighter ace was not enough adventure for one lifetime, Raymond Collishaw returned after the November Armistice to the RAF. In July 1919, he was given command of 47 Squadron and sent to Southern Russia to aid the White General Denikin against the Red Army. The RAF men carried out their duties with the White armies efficiently, but their numbers were too few to have any effect on the outcome. Collishaw organized a flight of Camels against the Red Army's Nieuports and even flew on fighter patrols over the Volga front himself. On 9 October he shot down a Albatross D-V — his 61st victory. Shortly afterwards, he contracted typhus and was hospitalized.

As late as March 1920, the Canadian carried out bombing missions for General Wrangel's White Army as it retreated from the Bolsheviks. But, by this date, London ordered the British Military Mission to cease participation in Russia and the RAF prepared to return to England.

Abandoning their aircraft, Collishaw and his men boarded a train bound for the Crimea where the Allied fleet could evacuate them away from the pursuing Bolsheviks. The retreat was the most frightening incident in the Canadian pilot's life. The train was overloaded with fleeing Czarist families and their possessions[1]. Typhus spread through the refugees who were loath to jettison their dead. Worse still, throughout the 500-mile journey, they were pursued by a Bolshevik armoured train. Collishaw, out of his element, was helpless.

The evacuation was constantly delayed as peasants sympathetic to the Bolsheviks, tore up the track or ambushed them when they stopped for water. Whenever they sighted the Bolsheviks behind them, more Czarist aristocrats and their families on board committed suicide rather than risk capture and torture. The RAF personnel bought precious time by tearing up the track behind them but at a village called Bolshoi Tomak, before they could do so, the Bolsheviks rammed their train with a runaway locomotive. After fierce fighting, laden with the dead, the battered train limped into a Crimean port and the Royal Navy. Collishaw admitted that more than anything in his life, either before or after, this episode terrified him, and he suffered from nightmares of the pursuing Bolsheviks for decades after.

Remaining in the RAF in the 1920s, the Canadian served aboard an aircraft carrier and later commanded the air station at Upper Heyford. By 1939, he was posted to Egypt, in charge of what would become the Desert Air Force. Although badly outnumbered by Italian and later German aircraft and considered a sideshow by politicians in London, the Desert Air Force under his leadership destroyed some 1,100 enemy aircraft. Made a Companion of the Order of the Bath, he was brought home in 1942 and promoted to Air Vice Marshal. His final command would be 14 Fighter Group in Scotland.

When he retired in 1943, Raymond Collishaw remained active in local civil defence organizations throughout the war. After it, he returned to his home province of British Columbia and became a partner in a mining operation. He died on September 9 1976, having lived a life that few had – or would ever – experience, nearly long enough to celebrate his 93rd birthday.

---

1 McCaffery, D., *Air Aces: The Lives of Twelve Canadian Fighter Pilots* (Toronto: Lorimer, 1990), p. 43.

Glenn Curtiss at the controls of the AEA's *Silver Dart.* Canadian aviation owes much to this New York motorcycle manufacturer. ( NAM 5315)

# CURTISS, GLENN HAMMOND

The inventor of the aileron, Glenn Curtiss was an American aircraft designer whose early career had a profound effect on Canadian aviation. Without him, there would have been few Canadian pilots in the initial years of the First World War. With Alexander Graham Bell and John A. McCurdy, Curtiss can be regarded as a founding father of powered flight in Canada.

Glenn Curtiss was born in Hammondsport, New York on May 21, 1878. Like Orville and Wilbur Wright, he was a bicycle mechanic. But unlike the Wrights, Curtiss built his own aircraft engines. In 1902, he opened a shop to make and sell motorcycle engines. Even then, he was aviation-minded and in 1904 he designed and built the engine for the early dirigible *California Arrow*.

Motorcycle engines were the most suitable means of power available to early aircraft manufacturers. They were light, commonplace, and durable. Curtiss soon acquired the reputation as the best motorcycle engine manufacturer in North America. Orders were multiplying and in the spring of 1907 the inventor Alexander Graham Bell visited his workshop in Hammondsport, to have an engine built. Curtiss heard how Bell with two young aviation enthusiasts, Casey Baldwin and John A. McCurdy, were conducting motorized kite experiments at the inventor's home in Baddeck, Nova Scotia.

As none of the Canadians had any experience with engines, Bell invited Curtiss to come to Baddeck and help them. This invitation was to have a profound effect on the aviation heritage of both the United States and Canada. Curtiss agreed and on October 1, 1907, the Aerial Experiment Association was formed.

All of the members worked that fall on Dr. Bell's tetrahedral kite *Cygnet*. On December 6 it was test-flown, carrying a person over the Bras d'Or Lake, after a small steam boat had pulled it into the air. After the experiment, the AEA decided to transfer to Curtiss's home at Hammondsport and attach his engine to their kite. Through 1908 they built and flew the *Hammondsport Glider* and in the early winter, the fully powered *Red Wing*.

On March 12, 1908, this latter machine was taken out onto the frozen surface of Lake Keuka, near Hammondsport. Curtiss had an air-cooled engine fitted to it, with each cylinder having its own carburettor. As if with a mind of its own, the *Red Wing* started up and moved across the ice on its own accord. Fortunately, the American and his workers were on skates and quickly caught up with it to put John McCurdy on board for the flight. McCurdy thus became the first Canadian and British subject to fly. Curtiss himself flew for the first time on May 22, 1908.

He then designed the *June Bug* a more solid aircraft which he flew 456 feet on June 6. That July the motorcycle manufacturer won the Scientific American Trophy for flying a heavier-than-air machine a full kilometre under test conditions. Throughout the busy summer days, Curtiss and McCurdy worked to improve the performance of the *June Bug* and build the next progression — the *Silver Dart*. In December, the new aircraft, powered by Curtiss's 35-hp. water-cooled engine, was test-flown.[1] Dr. Bell asked that the *Silver Dart* be shipped by rail to Baddeck where on February 23, 1909, John McCurdy made the first flight over Canadian soil.

At this stage, all the members of the Aerial Experiment Association were ready to go their own ways, and the scientific body was dissolved on March 31, 1909.

Curtiss remained at Hammondsport, developing and attempting to fly the *June Bug* off the lake. He soon knew that the conventional means of providing lateral controls, by warping the wing had reached its zenith. Wing warping had been invented by Orville and Wilbur Wright, to control the motion of their aircraft. Although accepted by the aviation world, it was awkward and limited in its application. After much experimenting, Curtiss developed the aileron. These were surfaces independent of the wings that offered better control. He mounted them on the trailing edges of the wings, setting the standard for ailerons forever. The Wrights claimed that their warping patent covered all means of lateral control on aircraft, and they took Curtiss to court. The litigation was settled in Curtiss's favour, and he was issued with the patent for the aileron on December 5, 1911. As if to vindicate him completely, the Aero Club of America awarded him Pilot's Licence No.1, with Orville and Wilbur earning licences No.4 and 5 respectively!

Curtiss continued to work on his seaplanes, gaining national attention in January 1911 by landing and taking off on the water at San Diego, California. By now, he was wealthy enough to indulge, full-time in his passion for aviation. His factory was turning out aircraft in a production line, just as Henry Ford was doing with automobiles. His friend from AEA days, John McCurdy, used a Curtiss aircraft to make the first flight across the ocean — from Key

West, Florida, to Cuba. But it was the First World War that provided the real boost to aircraft development, and especially for Curtiss.

Military aeronautics had begun in Europe as early as the Franco-Prussian War, when balloons were used for artillery observation. By 1910, Britain, France, and Germany had formed their own flying corps, and at the outbreak of the First World War were able to field aircraft for reconnaissance. By 1914, Allied High Command and the public accepted that the aeroplane was a potential weapon. The speed of aerial warfare rapidly accelerated, from pilots taking potshots at each other with hunting rifles to using fixed machine-guns by the end of 1914. In London and Ottawa, the mass production of aircraft and training of pilots became major concerns.

In May 1915, with America neutral, Curtiss opened his aviation school at Long Branch near Toronto. At the same time under the management of John McCurdy, Curtiss Aeroplanes & Motors began making aircraft at a factory on Strachan Avenue, Toronto. On the whole North American continent, only Curtiss possessed sufficient experience, trained personnel and capital to begin operations of this magnitude. In the spirit of Yankee entrepreneurship, it was an all-inclusive 'package deal.'

Two Curtiss F-flying boats were brought over from Buffalo, New York and hangers were erected at Hanlan's Point, on one of the islands in Toronto Harbour. Pupils were taught elementary flying on the F-flying boats and later progressed to wheeled aircraft at the Curtiss facility at Long Branch. The trainers used were the Curtiss JN-3s made on Strachan Avenue. The complete course took about 400 minutes in the air at a cost of a $1 a minute. Within a year, 66 pilots had graduated from the Curtiss school, going off to swell the ranks of the Royal Flying Corps. By the time the British government took over the whole operation in 1916, a total of 129 pilots had been trained.

Besides turning out the JN-3s, Curtiss Aeroplanes & Motors built the giant Canadas. These were lumbering, multi-engined bombers that had the distinction of being the first Canadian-designed military aircraft and the first ever to be exported. The prototype was test-flown at Long Branch in July, 1915, and shipped to England for trials. Only 12 Canadas were built, the aircraft becoming obsolete before they reached the frontlines.

In late 1916, when the British and Canadian governments decided to take control of the aviation industry, Curtiss was bought out, and the Strachan Street plant closed. But Canadian Aeroplanes Ltd. chose the Curtiss JN-4 or Jenny to licence-build, by the hundreds for the Royal Flying Corps. Curtiss, the former motorcycle engine manufacturer, still provided the engines, the 90-hp OX-5s.

By 1918, with the United States now embroiled in the conflict, the U.S. Navy realized the need for long-range flying boats to counter the submarine

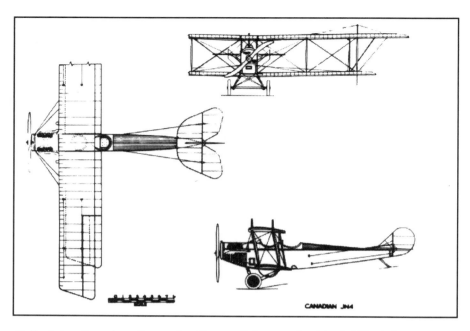

Called the *Canuck* or *Jenny*, the Curtiss JN4 was the Model T of the aviation industry. (NAM18498)

threat. Curtiss manufactured his HS-2L flying boats, the first long-range patrol aircraft, for convoy surveillance. After the war they would come into their own and become the mainstay of the forestry protection services in Quebec and Ontario.

But the aviator who had flown the kite-like *June Bug* on Lake Keuka only 10 years before had more daring schemes than the HS-2L in mind. Using his experience with seaplanes, Curtiss designed the Navy-Curtiss (NC)4.to fly the Atlantic. Powered by four 400-hp Liberty engines, the machines weighed 14 tons each and had a crew of five. Too late for the war, the four NC aircraft were chosen to attempt the first trans-Atlantic crossing from Far Rockaway, New York, via Newfoundland and the Azores to Plymouth, England. Three of the Curtiss flying boats took off on May 8, 1919, with the sole survivor landing at Plymouth on May 31.

Although Glenn Curtiss died in Buffalo, New York on July 23, 1930, the Canadian connection with his company continued through the century. In 1928, the Curtiss company purchased a controlling interest in Reid Aircraft of Montreal to turn out bush planes. During the Second World War, Canadian

Car & Foundry built Curtiss Helldivers and the Curtiss P-40 Kittyhawk was used by several home-front RCAF squadrons.

The inventor of the aileron, Glen Curtiss must also be honoured for the deep and critical relationship he had with early Canadian aviation.

---

1 After the *Silver Dart* crashed, Curtiss's engine was used to power a motor launch. Eventually, it too, broke up, and the engine lay on a river bed for years before its historical significance was realized. Today, it has been refurbished and is part of the collection at the National Aviation Museum.

C. H. 'Punch' Dickins, Edmonton, Alberta, 1930.   (NAM 1238)

# DICKINS, CLENNEL H. (PUNCH)

In a single year, 'Punch' Dickins flew 1,035 hours, covering some 87,467 miles — a formidable achievement even today, but more so in the 1920s. In keeping with his career, much of this was from airstrips in the far North, over featureless land with no amenities.

Clennell Haggerston Dickins, known in aviation circles as 'Punch', was born in Port la Prairie, Manitoba, on January 12, 1899. The family moved to Edmonton where, in 1917, Punch joined the 196th Western Universities' Battalion of the Canadian Expeditionary Force. He transferred to the Royal Flying Corps and joined 211 Squadron in France in 1918. By the time the war ended, Dickins had shot down a total of seven enemy aircraft and had been awarded the Distinguished Flying Cross. He returned to Canada and became one of the original officers of the Canadian Air Force in 1921.

Posted by the newly formed RCAF to Edmonton, Alberta, Dickins was made responsible for the testing of its British-made Siskin fighters, in winter conditions. He experimented with methods of keeping their Jaguar engines heated, and fitted skis to the wheeled aircraft. He also carried out extensive photographic work, even being on hand to record the famous cross country flight of McKee and Godfrey in 1926. With his experience in flying in adverse conditions in the North, it wasn't long before Dickins left the RCAF to join Western Canada Airways (WCA) and fly mail and prospectors.

In August 1928, Dominion Explorers of Toronto sponsored a mining exploratory flight over the unmapped Barrens, a vast area between Baker Lake and Stoney Rapids. In the hope that it one day could be opened up to large-scale mining, they chartered Dickins with his Fokker G-CASK to fly their team over the territory.

This was the first time such a flight had been attempted. With an absence of landmarks to navigate by or even outposts to land at should help be required, without radios or even recent maps, the party set off from Winnipeg. Punch flew along the Nelson River to Jackfish Island, then to Churchill on Hudson's Bay, to Mistake Bay, to Chesterfield Inlet, and west to

Baker Lake. The territory was a mining company's paradise as the rocks clearly held great quantities of mineral wealth. The pilot later recalled that they rarely saw a living thing once past the tree line. The party covered more than 4,000 miles in less than two weeks, a feat that would have taken two years by foot or canoe. In recognition of this and other exploratory flights, Dickins was awarded the Trans-Canada Trophy in 1929.

Soon engaged in running an airmail service from Edmonton down the Mackenzie River to Aklavik, Northwest Territories, WCA was contracted by the Post Office to carry the mail to Fort Simpson. Up to this time, as all mail was taken from the railhead by dog team, only letters could be posted. Using Fokker Super Universals, Dickins and his fellow pilots began regular mail flights through the Northwest Territories.

Before the use of radio, a pilot had to know how to navigate by the sun alone. Because of the proximity of the North Magnetic Pole, the compass was erratic and the shifting ice fields made what maps did exist useless. Often the only landmarks a pilot had by which to navigate were the ever-present ice-pattern ridges and the wind-sculpted Arctic topography. There was no meteorological organization to help with weather forecasts nor convenient alternate airports with fuel supplies if the destination was fogged in. It took a special breed to fly in that unforgiving land.

Dickins was the first to deliver furs by air — flying the valuable cargo from the trappers to the auction houses in Winnipeg. He became the first pilot to cross the Arctic Circle on the Mackenzie River. In the territory he flew, few natives had even seen an aircraft. Fewer still, he later recorded, believed that his machine could fly.

Most Inuit, WCA pilots realized, had never seen an automobile or a steam locomotive, but with the spread of aviation in the Arctic, native people began to take aircraft for granted. The story that Inuit began using aircraft for letter paper has some truth. They would pick out a light-coloured aircraft and scratch on its fabric with knife or pencil:

HELLO FROM THE DIOMEDE PEOPLE
WE GOT TWO WHALES THIS YEAR
PLENTY WALRUS
OUR PEOPLE ARE ALL WELL[1]

Messages like this would travel on the aircrafts' sides like a flying billboard for thousands of miles, to be enjoyed and added to. The pilots however were reportedly not too impressed with the holes that the scribbling made on their fabric.

On his second flight to Aklavik, Dickins and his co-pilot crash-landed on rough ice. Both the undercarriage and propeller were smashed, and the prospects of being found were slim. With great ingenuity, both began rebuilding their landing gear from pipes in the cargo hold. They then hammered out the propeller and cut eight inches off each tip. It took five days to make the repairs, but they were able to fly the Fokker home.

Again in 1929, Dickins made another pioneering flight to Great Bear Lake to rescue Gilbert LaBine, the man who discovered pitchblende, the ore that contains radium and uranium. When the prospecting expedition commanded by Colonel C.H.D. MacApline, the president of Dominion Explorers, disappeared in September, Punch Dickins was hired with other bush pilots to find them. Almost the day before the approaching winter made flying impossible, a wireless operator at the Hudson's Bay Company post at Cambridge Bay sent word that the expedition had arrived, tired and starving, but safe. But the search had taken its toll on the bush pilot community. Some 29,144 miles had been covered in freezing conditions. Aircraft had been lost, a crew member's toes had been amputated, and there were times when the rescuers themselves had to be rescued.

Western Canada Airways was taken over by Canadian Airways in 1930 and Punch Dickins made the superintendent of the Mackenzie District. Credit must be given to its president, James Richardson, for securing the lucrative government contract to fly a nightly Prairie Mail Service from Winnipeg to Edmonton in 1930. Once the night-landing facilities were in place along the route, Punch continued to do much of the flying. It came as a shock to him and the airline when, because of the Depression, Ottawa cancelled the nightly service in 1932.

Through the '30s, Dickins continued to fly across the Yukon and Northwest Territories. As mundane as this might seem, it took a special type of pilot to do so. With airstrips usually a few hundred meters of gravel beach or a hillside with drainage, a bush pilot had to be very lucky or foolhardy to set his wheels down in the Arctic. Like all experienced pilots, Punch knew to stay clear of invitingly flat areas in the frozen tundra as they became swampy quagmires with a sudden rise in temperature. His undercarriage had to be constantly checked and tires under-inflated to spread the weight of the aircraft as it bounced on a thin crust of earth in the summer or new ice in the early winter and late spring. Good pilots could tell from the type of vegetation if there was drainage on a landing site in the summer and what sort of load they could get in and out with.

In tribute to his reputation, federal ministers asked that Dickins fly them on mineral inspection trips over the North. In 1935, Punch Dickins was made an Officer of the British Empire. That same year, he was promoted to general superintendent of Northern Operations for Canadian Airways.

During the Second World War, because of his experience in long distance flying, Punch was appointed manager of Canadian Pacific Railway Company (Air Services) to run the delivery of aircraft from North America across the Atlantic. In the post-war years, he joined the de Havilland Aircraft Company, becoming responsible for marketing their Beaver and Otter aircraft. In 1954, in recognition for his pioneering flights, the province of Alberta named a lake after him. In 1967, Dickins became an Honourary Fellow of the Canadian Aeronautics and Space Institute — the highest honour the Institute can bestow.

---

1 Potter, Jean, *The Flying North* (New York: Macmillan, 1947), p. 188.

# ELLIS, FRANK H.

One of the original 'Early Birds,' Frank Ellis retired from flying in the 1920s. For the next six decades he devoted himself to writing about the pioneer era in Canadian aviation that he had been part of. His first-hand accounts of events and personalities in those far off-days have ensured that they are not lost to future aviation historians.

Frank Ellis was born in Nottingham, England in 1893. He recalled that his love for all things aeronautical began in 1911 when he collected sets of Player's cigarette cards showing what were called aerial machines. As did a lot of air-minded boys of the period, he joined kite clubs, and because his mother was a milliner, he had endless supplies of thread with which to build them. After graduation, Ellis was employed by the Raleigh Cycle Company, which was then beginning to branch out into manufacturing motorcycle engines. Just before the family emigrated to Calgary, Alberta, in 1913, the apprentice bicycle-maker won a prize for his model of a Bleriot aircraft.

In Calgary, Ellis met Tom Blakely, a kindred soul, and the two set about rebuilding an abandoned Curtiss biplane. The Ellis-Blakely machine was christened *West Wind,* and on June 25, 1914, took to the air from some flat land on the city's outskirts. To be up at dawn and begin their daily flying experiments before commuting to city jobs, the two youths slept by their aircraft in a pup tent. Barely holding onto their employment, both flew all through the summer becoming progressively more daring — and dirtier, as all water had to be carried to the campsite from a spring a mile away.

On August 4, 1914, when news of the war reached Calgary, Ellis and Blakely gallantly offered their services in succession as trained pilots to the Canadian, British and French governments. They received a polite refusal by telegram from Ottawa: the Canadian military did not anticipate any use for aviators in the present conflict. The British and French were more realistic: they asked the Canadians to present themselves at the airship offices at Wormwood Scrubs and Rouen immediately!

Frank Ellis at the controls of the Ellis-Blakely *West Wind* at Bowness Park, near Calgary, 1914. (NAM 26007)

However, through 1915 they continued their flying at Shouldice Park, then many miles west of Calgary city limits. But constant exposure to the prairie weather and the rough handling in the experimental flying techniques made the biplane age rapidly, and it began to fall apart. The seams were soon beyond repair and the fragile undercarriage could not sustain the heavy landings much longer. Towards the end of the summer, while it was tethered to the ground, the *West Wind* was carried aloft by a gust of wind and deposited a half mile away. When Ellis and Blakely reached it, it lay in a tangled mess, beyond repair.

With the war over, hundreds of surplus Curtiss aircraft became available and Ellis embarked on a career as a barnstormer. In an age when few had seen an aircraft in flight, let alone been aloft, the pilot was the combination of a circus acrobat and gladiator. The public paid to see death-defying stunts performed before their very eyes. It did not matter that the wind was not right, or that the landing area was too short. Rather than disappoint an audience who would quickly begin to demand their money back, the pilot(s) would be forced to take off and perform suicidal stunts for a pitiful fee. Spectacles of sitting astride the fuselage behind the cockpit soon gave way to wing walking, which was then surpassed by hanging from the undercarriage of the aircraft. Those heady days are catalogued in Ellis's book *Canada's Flying Heritage*. Indeed, the public's hunger for blood and thrills led to so many deaths among the airmen that in late 1928 both the Canadian and American governments brought in regulations to curb the slaughter.

Frank Ellis barnstormed through the prairies as a mechanic and later a pilot in an Avro 504. As he was later to write, he was "of an age when danger did not enter his thoughts." In 1919 he was employed by Allied Aeroplanes in Toronto to look after a Curtiss Jenny based at the Ontario resort of Crystal Beach. His company worked the dollar-a-minute trade where holidaymakers were enticed to be taken aloft at that rate. That summer Frank was to be involved in an exploit that ensured his place in the record books.

Although the first parachute jump in Canada had taken place in Vancouver seven years before, it had been made by an American. That summer the Irvin Parachute Company of Buffalo, was experimenting with its 'backpack'— a rudimentary form of parachute. On July 4, the inventor, Leslie Irvin, jumped out of the AAL Jenny at a height of 2,000 feet into Lake Erie. Irvin left the wet parachute at Crystal Beach, and the next day Ellis thought that he would try it out.

It was a heavy linen contraption that came with a large inflated inner-tube, and he had difficulty fitting into the rear seat of the aircraft with it on. The pilot took the Curtiss up to 1,800 feet over the lake, and as soon as they saw a boat below (and were thus assured of rescue), Frank clambered out of the rear cockpit onto the wing. Then the pilot put the aircraft into a steep climb and Ellis jumped off. The parachute worked "like a bag of cement" and he plunged deep into the water before the inner tube carried him to the surface. The linen thread used in the 'chute' shrank so quickly in the water that Frank was almost strangled by the straps. But the occupants of the boat had seen him and rushed to his rescue.

Ellis became the first Canadian and British subject to make a parachute jump. The red welts left on his body by the tightened linen straps ensured that silk would be used in future parachutes. Less than a month after his jump, parachutes would be used in an emergency situation.

Although Ellis remained proud of his exploits in early aviation through-out his life, his greater contribution was in the field of aviation history, and even archeology. When he retired from active flying in the late 1920s and worked in Vancouver as a bus driver, he poured all his energies and spare time into collecting, researching, and publishing material on Canadian aviation.

He doggedly tracked down aviation pioneer William Gibson's original air-craft motor — the first Canadian aircraft engine. He hunted for and found one of the two home-made propellers that were built at Fort Simpson and used to fly the Imperial Oil Junkers 500 miles to Peace River Crossing in 1921. Aware of their value, Ellis donated these precious items of our aviation heritage to the National Research Council in Ottawa.

But perhaps his greatest contribution to this nation is his classic study *Canada's Flying Heritage*. More than any other 'Early Bird', Ellis realised that if he did not set down the trials and adventures of pioneering Canadian aviators for posterity, they would be lost forever. Every subsequent aviation historian has poured over this comprehensive collection of material, all the time realising that it was written by an aviator who is part of history himself.

Ellis wrote two other books on Canadian aviation — *In Canadian Skies* in 1959 and *Atlantic Air Conquest* in 1976. The last was co-authored with his wife, who predeceased him in 1977. In October 1972, through the efforts of the Canadian Air Historical Society, Ellis was awarded the Medal of Service of the Order of Canada, and in 1974, the Government of Manitoba named Ellis Bay in his honour.

He was the last survivor of the five original Canadian members of the 'Early Birds,' that distinguished society of pilots who flew before the First World War. The other Canadians were John McCurdy, Casey Baldwin, William Stark, and William Gibson. They were the catalysts of Canadian aviation.

Frank Ellis died in Vancouver on July 4, 1979, one day before the sixtieth anniversary of his parachute jump in 1919 at Crystal Beach, Ontario. His memories, within the covers of *Canada's Flying Heritage*, of how the country shrank within his lifetime, are one of this nation's treasures.

Walt Fowler in his Trans-Canada Air Lines uniform, Moncton, 1938.
( NAM 18764)

# FOWLER, WALTER Warren (WALT)

Sixty years ago, to mention casually that you had travelled in an aircraft was akin to taking the Concorde today. Apart from air force personnel and barnstormers, few people had actually gone aloft. Taking a passenger-carrying aircraft then was uncomfortable, noisy — and dangerous. The press delighted in lurid photographs of air crashes and crumpled bodies. Airports were little better than muddy fields situated as far from a city centre as possible. Why, most people wondered, should one give up the luxury of a Pullman railcar that wafted its passengers from a grandiose, downtown railway terminal to another for a flight? Canadian airlines then were in their infancy and far-sighted men like Walt Fowler were attempting to improve their reliability and public image.

Walter Warren 'Walt' Fowler was born in Sackville, New Brunswick, on September 8, 1906. He earned a commercial pilot's licence at Jack Elliott's Flying School in Hamilton, Ontario in 1928. The school's owner was a colourful character who had come into the aviation world quite by chance. Elliott had been in the music business for many years and had received two crated war-surplus Curtiss Jennies as payment on a debt. He had heard that it was the policy at that time for the Canadian Air Force to teach anyone who owned an aircraft to fly, and Elliott not only took advantage of this but purchased several more Jennies to open his own flying school. Charging a dollar a minute for instruction, he had his students dress in a military uniform, down to the wearing of puttees, and operated from the ice off Burlington Beach, Hamilton, in the winter. The young Fowler learned to fly in good company as many of the school's graduates went on to run their own airlines and one classmate, Frank Young, would be awarded the Trans-Canada Trophy 30 years later.

International Airways of St. Laurent, Quebec, hired Walt Fowler in 1929 as a mechanic and general assistant at their Windsor, Ontario, base. Here,

Fowler taught flying to build up his hours, and worked as a night mechanic at the Windsor Flying Club.

In 1929, International Airways merged with Canadian Airways, and he was sent to their offices, first in Detroit, Michigan, and later Prince Edward Island to fly as an airmail pilot. After a short course on instrument flying at Camp Borden with the RCAF, Fowler took charge of the Maritime Region for Canadian Airways, based at Charlottetown, P.E.I. To improve the airline's public image, he began daily flights between the Maritimes and Montreal. This was especially difficult as it involved taking off whatever the visibility. By 1930, night runs were added as well.

Passengers and the post office were loath to use aircraft that could only fly during daylight hours when overnight trains were available. The airlines knew that, if they were to be taken seriously, they would have to be more than a clear weather, daylight form of transport.

But there were no blind-flying instruments, radio communications, or trustworthy maps. With their sole navigational instrument a compass, pilots had, up until the late '20s, only flown when they could see the ground. Even Louis Bleriot, before he took off on his famous cross-Channel flight, is reputed to have first asked a spectator to point the way to England. He then followed the cross-Channel ferry.

During the First World War, luckless pilots became so lost that they would sometimes land at enemy airfields by mistake or drift until the fuel ran out and be forced down in the nearest cow pasture. After the war, fortunate pilots followed the 'Iron Compass'(main railway line) between two cities. This was hardly popular with the plane's passengers, who could see the train going faster than they were.

Experienced pilots knew that flying in clouds for any length of time led to loss of control and invited a potentially disastrous spiral. They had no idea about the winds aloft, or how high the cloud ceiling was, or if freezing rain would ice up their flaps and ailerons. Canadian bush pilots flew along the winding rivers not only because they provided a safe haven on which to land but also because they were certain as to the course.

In a country as vast as Canada, flying by night before the '40s was a dangerous experiment. Then, few areas were lit outside the main cities. There was little rural electricity and infrequent car headlights lit country roads. A pilot taking off on a moonless night was at the mercy of the weather and his compass, unable to radio or even verify his position with landmarks. Helpful farmers sometimes lit bonfires on their properties but this could not be relied on.

As early as 1925, the United States government had extended a lighted airway from coast to coast. It consisted of 18 lit terminals, 89 emergency airfields, and 508 rotating beacons. As the former night-mail pilot Charles Lindbergh was to prove, long-distance night flying was well advanced with the United States Mail Service. In the best dynamic tradition of American business, entrepreneurs like Donald Douglas, William Boeing, and Juan Trippe pushed Congress into authorizing money and schemes for a trans-continental air-mail route.

On the Canadian side of the border, the rare night-flying pilot still counted on occasional bonfires and car or locomotive headlights to guide him. In 1930, when the certain air routes (e.g. between Winnipeg and Calgary) became night runs to fly priority mail, the Canadian government had beacon lights built every 10 miles, and lighted emergency airfields were established. At strategic points, rotating searchlights flashed an identifying code, and powerful floodlights were installed at airport terminals.

While the lighted airways were an improvement over blind flying, weather could still ground all aircraft. Worse, in 1932 as a result of the Depression when the post office could no longer afford them, even the night mail services were discontinued. Prime Minister R.B.Bennett reasoned that there would be little gratification for the 300,000 Canadians then on relief to see a government-sponsored aircraft fly over every night.

It was instrument and radio flying that finally emancipated the pilot and allowed airlines to operate not only at night but in poor weather. The first low frequency radio-ranges were commissioned to transmit Morse code letters to opposite quadrants of the pilot's receiver. Where they joined, the two signals meshed in a steady hum and this was the course that the pilot followed. Walt Fowler had earned his air engineer's licence in 1935 and was appointed to run the first civilian instrument flying course given by the RCAF. By 1937, he was flying out of Senneterre, Quebec, still for Canadian Airways. His routes ranged far over northern Ontario and Quebec.

That same year an Act of Parliament was passed that would have far reaching effects not only for Fowler but for the future of the country. On April 10, 1937, Parliament assented to the Trans-Canada Airline Act. Rather than award the contract to any of the existing airlines, it opted to create a wholly new national company to carry passengers, mail, and freight across the country. The minister for transportation, C.D. Howe sponsored Bill No.74 to create Trans-Canada Air Lines, and he told the House of Commons that $7 million had been allocated for the construction of landing fields, radio equipment and hangers. Like the railway before it, the airline was to be the glue by which the nation was held together. The Department of Transport, Howe said would be responsible for the provision, maintenance, and operation of emergency landing fields, lighting, radio range equipment, and meteorological services.

There was much criticism from the Opposition benches that the new airline's directors chose an American, Philip Johnson, to become vice president of operations. But to Walt Fowler, this must have been of little consequence, as he was picked to be one of the first pilots in the new airline.

At first, Trans-Canada Air Lines used exclusively Lockheed products — Lodestars and Electras. These last were the preferred aircraft of trail-blazing pilots such as Amelia Earhart, and perfect for the image of the new airline. On July 30, 1937, TCA flew C.D. Howe himself from Montreal to Vancouver in an Electra on an inspection flight of the airway. By the time the regular Montreal-Vancouver scheduled flights began in late 1938, Fowler had already familiarized himself with every TCA route and aircraft. In 1942, he inaugurated the Moncton-St. John's, Newfoundland, service. He was now flight superintendent of the Atlantic region and had earned the title 'Mr. TCA'.

As an authority on instrument flying, he was ordered by the government to organize TCA`s aircraft ferry service during the Second World War. This was an exercise that would have been impossible a few years previously. With its facilities and personnel, TCA co-ordinated the routing of hundreds of bombers to be flown to Canada from their American manufacturers for overseas shipment. Fowler`s instrument-flying techniques were taught to the pilots of the Atlantic Ferry Command who took the aircraft on to Europe.

The airline's rapid expansion in the post-war, years with the acquisition of the Canadair North Star, meant more senior appointments for Fowler until he became general manager for the whole of the Maritimes in 1969.

When he retired in 1971, he had logged 10,000 command hours in 41 different types of aircraft from Curtiss JN-4s to jet airliners, without a single crash. Fowler had come a long way from Jack Elliott's Flying School.

# GILBERT, WALTER EDWIN

From 1917 to 1949, Walter Gilbert was deeply involved in the development of aviation in the Canadian northland. Through his surveys and mapping work, he helped record the topography of the North.

Born at Cardinal, Ontario, on March 8, 1899, Gilbert was taken, as a boy, in 1910, to the first Canadian aerial meet. It was held on June 27 at the Lakeside Race Track near Dorval, Quebec. John Bassett, a junior reporter for the Montreal newspaper the *Gazette*, in a fine display of entrepreneurial skills, publicized the event sufficiently to attract crowds of spectators and prominent aviators. One of them was Count Jacques de Lesseps, son of the builder of the Suez Canal, another was Walter Brookins, a protégé of the Wright brothers. The first flight over a Canadian city was made when de Lesseps flew for 45 minutes in a wide circuit over Montreal. The *Gazette* enthused that at one time three aeroplanes were simultaneously in the air!

No doubt this is where Gilbert decided to become an aviator. He joined the Royal Flying Corps at the age of 18 and was trained at Camp Borden in the summer of 1917. Although he was posted to an operational squadron in France, he did not see any fighting and was demobilized in 1919.

He went west to Vancouver in 1922 and did some survey work in the mines. Gilbert received his commission as a flying officer in the RCAF in 1927 and was employed as a pilot on forestry patrols in northern Manitoba. He was a proficient map maker, and as a member of the British Columbia Mountaineering Club he enjoyed sketching the terrain, filling in the details while flying over it.

Flying itself was still a bare-subsistence profession then. With the railway in its heyday, aerial travel was an expensive mode of transport. The aircraft were too primitive to carry many passengers, and the general public was too afraid to fly. Most pilots lived a hand-to-mouth existence, dreaming of working for a air charter company.

Gilbert joined Western Canada Airways in 1928 as a commercial pilot. This was a prized concern then, for it offered regular employment and

W. E. Gilbert (left) and Stan Knight (right) in a typical winter setting.
(NAM 2129)

possessed relatively modern aircraft. Western Canada operated four Boeing B1E flying boats doing fisheries patrols for the government on the coast between Vancouver and Prince Rupert. It later added two Fokker Super Universals to its fleet.

But its flagship aircraft was a sturdy, single-engined Junkers W-34, CF-ABK. Now part of a diorama in the National Aviation Museum, this Junkers became a legend among the bush operators in the 1930s. The Treaty of Versailles had prevented the Germans from manu-facturing military aircraft, so they turned their skills to passenger airplanes.

W.E. Gilbert. (NAM 1236)

Hugo Junkers built the F-13 series, rugged single-engined aircraft, which were made out of corrugated duralumin. The Junkers W-34 was even larger than the F-13, and remained through the next two decades as one of the workhors-es of the air.

Gilbert flew the new Junkers to Stewart, British Columbia, on contract for the Consolidated Mining and Smelting Company in 1929. He was trans-ferred to Fort McMurray, Alberta, in 1930 and it was from there that he and Punch Dickins made several historic flights into the Mackenzie River area. On May 16, 1930, they both flew the Fokker Super Universal G-CASK into Great Bear Lake where the greatest mineral strike of radium in the north had just taken place.

The following year, Gilbert gained much valuable experience flying to Aklavik, transporting the Anglican bishop on an inspection tour of the far northern parishes. On his return, Western Canada Airways was asked by Ottawa to fly Major Lauchie T. Burwash, a government explorer, to King William Island, in search of remains of the ill-fated Franklin expedition.

A British explorer, Sir John Franklin had left England with two ships in 1845 to try to find the Northwest Passage. The whole expedition disappeared, but reports of their remains had been made periodically. The federal authorities also wanted to photograph the Arctic coast and the area around the magnetic pole. Gilbert was chosen as pilot and a Fokker Super Universal was used.

Major Burwash was flown to the mouth of the Coppermine River on the Arctic coast. On the way, Gilbert assisted in salvaging another Western Canada Airways Fokker G-CASK that had been abandoned at Dease Point during the infamous MacApline expedition. This was flown back to Fort McMurray, to the relief of Western Canada Airways. Burwash's work was soon completed, and Gilbert prepared to fly out of Coppermine when disaster struck.

It appeared in the form of a burnt-out piston as he was about to take off, and it looked like the party would be marooned there until the spring. Fortunately, the annual supply boat had just left Coppermine, and Gilbert was able to race after and catch up with it to send a radio message. The replacement aircraft that his company sent was none other than G-CASK, the Fokker that had just been found. In it, they left Coppermine and flew over the outer fringe of the Great Polar Ice Pack, reaching the North Pole on September 6, 1930.

On their return, they landed on a lake on King William Island and looked for campsites and graves of the Franklin Expedition, but little was found. With the onslaught of winter, the ice was starting to form on the edges of the lake, and Gilbert had a difficult time taking off. It was starting to snow heavily by then and the Fokker sped south to Edmonton which was reached on September 29, 1930. A total of 5,000 miles had been covered.

For his flying, Gilbert was honoured with a Fellowship in the Royal Geographical Society in 1932, and in 1934 he was awarded the Trans-Canada Trophy.

He continued to fly on freight and mail runs, transporting mining equipment, letters and sometimes even passengers between Aklavik, Yellowknife River, Fort Good Hope, Hay River, Fort Simpson, and Arctic Red River. The airline's mainstay in the Depression was the shipment of furs. Gilbert would write about his flying experiences in his book *Arctic Pilot*. These first-hand accounts of pioneering aviation in the far North are an invaluable record.

In 1937, he was appointed superintendent of Canadian Airways, and in the following year was transferred to Vancouver when the company became part of Canadian Pacific Air Lines. After the war, he left Canadian Pacific to form his own airline, Central British Columbia Airways. It was unsuccessful and he left aviation in 1949 to open a fishing resort at Chilliwack, British Columbia.

# GODFREY, ALBERT EARL

S panning a long and eventful career — first with the Royal Flying Corps and later the RCAF — Albert Godfrey rose from the ranks to become an Air Vice Marshal. His enthusiasm for the innovative was always in the forefront.

Albert Earl Godfrey was born on July 27, 1890, in Killarney, Manitoba. Soon after his birth, his parents moved to Vancouver. His interest in aviation began in 1911 when he saw Billy Stark and Bill Templeton, two of the 'early birds' in British Columbian aviation, barnstorm at the local fair. He took up an interest in motorcycle racing and thrice became the provincial champion — in 1911, 1912 and 1913.

When the First World War broke out, Godfrey joined the 11th Canadian Mounted Rifles and went overseas. However, every attempt he made to join the Royal Flying Corps was refused by his commanding officer. At the Battle of Ypres, Godfrey realized that the determined German attacks could only be countered if the Allies were armed with automatic weapons. Always good with mechanical devices, he set to work to install a mechanism in his Ross rifle that converted it into an automatic weapon. His commanding officer was so impressed with this that he had Godfrey demonstrate it before General Currie, who offered to send the plans for the automatic rifle on to London. As they shook hands, Currie asked the young soldier if there was anything that he could do for him. Thinking quickly, and within earshot of his commanding officer, Godfrey asked for a transfer to the Royal Flying Corps. Within a week, his wish was granted.

By 1916, Godfrey was flying on bombing missions across German lines with No. 10 Squadron. He later joined No. 25 Squadron, and in a F.E.2b, shot down his first enemy aircraft. But it was in No. 40 Squadron then operating French Nieuports where Godfrey really excelled. He was soon credited with shooting down 17 enemy aircraft and two balloons, and he was awarded the Military Cross. Years later, he recalled that one of the pilots on his patrol was the British fighter ace Major E. "Mick" Mannock, who shot down 50 German aircraft and won the Victoria Cross.

Godfrey was posted back to Canada and promoted to squadron leader and he commanded the RAF training establishment at Beamsville, Ontario. At the war's end, he entered the civilian Government Air Operations Branch and flew fisheries patrols along the Pacific coast.

Air Marshall A. E. Godfrey. (NAM 11867)

In 1922, he returned to the Canadian Air Force to become commanding officer of Camp Borden and in 1926 the Superintendent of Air Operations in Ottawa.

But Godfrey is best remembered for is his cross-country flight with Dalzell McKee in the fall of 1926. An American millionaire, McKee was flying his Douglas float-plane to California to have it modified at their plant. With the idea of flying via northern Canada he asked the RCAF if one of their officers might accompany him. To the military, this was an excellent opportunity, with little cost to the taxpayer, to map a trans-Canada route for float-planes. As Godfrey was already ordered to inspect the RCAF base at Vancouver, he was chosen to act as co-pilot and navigator.

An adventure of this magnitude required much planning. As this was the first float-plane flight across Canada, gasoline had to be shipped out to specific lakes or river points across the country, RCAF seaplane bases alerted to expect the pair, and aerial maps had to be fully studied. The Douglas had a water-cooled engine that would have to be drained nightly so that it did not freeze in the northern latitudes. A spare engine was located, in case this happened.

In Godfrey, the RCAF had chosen well. The trip, much over still-unexplored country, was made in 34 hours and 41 minutes flying time, and the young airman never forgot the excitement of it.

The following year, McKee returned to Canada, and he discussed flying with Godfrey to the Mackenzie River and Alaska. This was, after all, the era of glorious odysseys by aeroplane. Sir Alan Cobham had just flown an epic round-trip of London-Australia-London in 90 days. The polar explorers Roald Amundsen, Umberto Nobile, and Richard E. Byrd were each searching for an undiscovered land mass above Canada. However much anyone protested, personal and national prestige went hand in hand with such flights.

McKee and Godfrey chose to use a Vickers Vedette, with which only Godfrey had some experience. McKee flew the aircraft to Lac la Pêche, Quebec, on June 9 to practise landings, and Godfrey went home to Ottawa to pack for the flight. Hardly had he entered his house when the phone rang. He was told that his partner, Dal McKee, attempting to land the Vedette on the lake, had been killed. Both Godfrey and Canadian aviation lost a good friend that day.

Undaunted, Godfrey planned another epic cross- country flight. This time he obtained a float-plane from Sherman Fairchild of the Fairchild Aircraft Corporation. Carrying mail and a co-pilot, on September 5, 1928, Godfrey flew from Ottawa to Vancouver in three days, with a flying time of 32 hours. Had the weather been good, he later said, the trip could have been made in two days, flying in daylight.

At Kamloops, British Columbia, they had run into the smoke from a huge forest fire, and visibility was nil. They tried to climb over the pall, but when this was not possible they were forced to bring the aircraft down to 1,000 feet and fly along the Fraser River. This was only slightly less hazardous: the river swept through narrow canyons, and the Fairchild had to be banked steeply to avoid hitting the walls.

At Vancouver, they planned a return journey by a more northerly route, and they fitted two other officers into the aircraft with them. But smoke from forest fires, this time in the Peace River area made visibility impossible, and Godfrey had to come down low over the river. This time, his luck had run out, and he crashed into the water. The party was lost for 12 days, sheltering in a trapper's cabin. The shroud of smoke that hung over the Peace River district made all rescue attempts impossible. Finally, one evening they heard the sound of a passing motorboat, and they lit a huge bonfire on the shore to get the operator's attention.

They were rescued and after a series of misfortunes — engine troubles and running aground they reached Peace River Crossing, and the welcome news was flashed to a worried Ottawa.

In recognition of these flights, Godfrey was made wing commander, and in 1931 took charge of the air station at Rockcliffe and, later, Trenton. Just before the Second World War, he made an inspection of the Maginot Line in France to report on air bases around it. In September 1940, he was appointed aide-de-camp to the governor general, the Earl of Athlone. In June 1942, he was promoted to air vice marshal and took over the duties of deputy inspector general for Eastern Canada.

He made it a point to visit every RCAF base in Canada, Newfoundland, Labrador, and Alaska, giving praise and encouragement to the men in the vastly expanded RCAF. He often took over the controls of his VIP aircraft himself, and in 1943 while on an inspection tour of RCAF Gander became the first air vice marshal to go on patrol over the North Atlantic. In the spirit that had made the young soldier invent a device to change his rifle into a automatic weapon, Godfrey joined a Liberator bomber crew on a U-boat hunt. He flew the 15-hour patrol with the young men, even manning one of the gun turrets.

He retired from the RCAF in 1944 and he was presented with the Trans-Canada Trophy in 1977. We can only speculate about his thoughts when he received the trophy that had been donated to Canada so long ago by his friend Dal McKee. Godfrey died in 1982, having seen Billy Stark, the aviation pioneer, in 1911, and the space shuttle 70 years later.

# GRAHAM, STUART

B y 1918, the combatants had built over 200,000 aircraft exclusively for the Western Front. When the war ended that year, more than 26,000 pilots had been trained for the Royal Flying Corps alone. A year later, there was little future for either the aircraft or their pilots.

There were no mail or passenger services (as there would be after the Second World War) to absorb the many restless young fliers or the surplus aeroplanes into commercial use. A flamboyant few went into barnstorming — earning a precarious living by stunting above crowds at exhibitions. The thrill of defying death coupled with the glamour of being in the public's eye led some to attempt record-breaking flights — across the Atlantic, across the Pacific, or over the Poles. But for the most part, the many unemployed fliers tried to make their skills commercially viable with mundane jobs such as grindingly boring forestry patrols, aerial photography, and bush flying. Stuart Graham was one such pilot.

Graham was born at Boston, Massachusetts, on September 2, 1896, and educated in Truro, Nova Scotia; Windsor, Ontario; and Wolfville, Nova Scotia. He joined the Canadian Army on the outbreak of the First World War and served as a machine gunner on the Western Front. On recovering from wounds, he was able to transfer to the Royal Naval Air Service. Graham trained as a pilot, and he spent the remainder of the war on anti-U-boat patrols — and awarded the Air Force Cross for his efforts.

On his return to Canada in 1919, he heard that a group of pulp and paper companies in the St. Maurice Valley, Quebec, were interested in mounting aerial surveys of their timber resources. Ellwood Wilson, the chief forester of the Laurentide Company was enthusiastic about the use of flying boats for forest-fire detection, and with a financial subsidy from the Government of Quebec, he purchased a pair of HS2Ls.

These Curtiss flying boats had been used during the war by the United States Navy to patrol the shipping lanes outside Halifax. One of the pilots who gained experience in long distance flying with them was Richard E. Byrd, who later became the first to fly over the North Pole. After the Armistice, the flying boats were dismantled and stored at Dartmouth, Nova Scotia.

The St. Maurice Forest Protective Association hired Graham as its chief and only pilot, his first job being to fly the aircraft from their base at Dartmouth, Nova Scotia, to Grand'Mère, Quebec. By this act, Graham became the first national peacetime professional pilot and the first bush pilot in Canadian history. This was a flight of 645 miles, and Graham flew his engineer, William Kahre, and his own wife in his plane.

They left Darmouth in the first aircraft, named *La Vigilance*[1] at 2:25 p.m. on June 5, 1919, and despite a thick fog arrived at Saint John, New Brunswick, on schedule. The mayor of the city organized a tumultuous welcome — even taking them to the opera, and the trio were only able to depart for Eagle Lake, Maine, on the 7th. The weather had changed for the worse, and the next leg of the trip to Lake Temiscouta, Quebec, was uncomfortable and wet as the open-cockpit Curtiss bucked the high winds and rain. Because they carried a letter from the lieutenant governor of Nova Scotia to the premier of Quebec, this trip is regarded as the first airmail flight between the provinces of Nova Scotia and Quebec. Four days after they had set out, the three air travellers arrived at Lac à la Tortue and taxied up to the beach at Grand'Mère.

Bringing the next HS2L was not as easy. They were late on setting out from Dartmouth on June 21, and because it was so heavily loaded by the 80 gallons of gasoline, the Curtiss almost hit a low railway bridge on take off. Over the forests of Maine, the oil pressure began dropping, and Graham put the aircraft down on an isolated lake for repairs. Fortunately, there was an inhabited log cabin nearby where the aviators asked for shelter. The ragged, desperately poor, backwoods family that lived there had never seen anything like the goggled, leather-suited aliens before, nor an aircraft, and regarded their visitors with dumbfounded shock. The whole encounter, Graham later recalled, took place in the form of mime.

On the 23rd, after still more misadventures, Graham and his crew landed at Grand'Mère. Here, the HS2Ls were to become the backbone of the forestry service, able to land on the many lakes in the province, and requiring only rudimentary maintenance.

The Laurentide Company served as the training school for many future bush pilots such as Romeo Vachon, Tommy Thompson and Roy Maxwell. In 1920 Graham and his engineer, Walter Kahre, flew a forestry engineer into an isolated area of Quebec to stake the first mining claim by air.

But for a financially hard-pressed aviation company, nothing could match the security of flying the mail for the post office. It was a prize sought after by

Stuart Graham in 1936. An aerial photographer, he invented an automatic viewfinder control for aircraft cameras. (NAM 23470)

all early aviators, from John Alcock and Arthur Brown to Charles Lindbergh and Stuart Graham. In 1927 the Canadian post office put into operation an imaginative scheme of cutting international delivery time by transferring mailbags from a trans-Atlantic ocean liner to a flying boat, as far out in the St. Lawrence as possible. The first successful flight with a load of incoming mail was made on September 16 in a Canadian Airways HS2L flown by Stuart Graham. The Canadian Pacific liner *The Empress of Australia* was met, and 500 pounds of mail were transferred to the aircraft, to be flown on to Montreal. It was soon understood that the stormy St. Lawrence was no place to heave mail bags into fragile flying boats, and eventually the mail was carried by the pilot boat to the port of Rimouski and then transported onward by aircraft.

After leaving the forestry service, Graham worked first for the Curtiss Aeroplane and Motor Company and later for Canadian Vickers in Montreal. Towards the end of the decade, he served with the RCAF as a pilot, engaged mainly in aerial photography.

The war had demonstrated the value of aerial reconnaissance for map making, and in still-unexplored Canada the science of aerial photography had been adapted to know the country better. In the 1920s, the RCAF was the only organization with the capabilities of such a vast undertaking, and they set about using their HS2L's for this purpose.

In suitable weather, the photographic unit pilots flew along a line taking photographs that were sent to Ottawa to be developed. Difficulties encountered were distorted or inaccurate photographs when the aircraft was not level (as when climbing or descending), or when blown off course or when flying too low. The height of 5,000 feet was seen as the most suitable, but because of low cloud, this was rarely possible. The departments of Indian Affairs, and Marine and Fisheries, provincial governments, and universities all clamoured for aerial photographs for their own purposes, and the RCAF was thus heavily engaged. An amusing side effect of this was that sometimes the mere presence of a photographic unit in the area led several bootleg alcohol makers to hurriedly destroy their stills. The hardy Curtiss flying boats, built for a war that had ended a decade ago, soldiered on until they could be replaced by the Fairchild FC.2's, an aircraft designed specifically for aerial photography.

With these flights, Graham's fascination with this science grew, and he invented an automatic viewfinder control for cameras. He also successfully patented a sectional canoe that came apart to be stored onboard an aircraft — and a remote-control, landing direction indicator for use at airports.

He left the RCAF in 1928 to join the Air Services Division of the government as one of two district inspectors for all of the country. Although his primary job was to test pilots for their licence, he also investigated air accidents and took part in searches for lost airmen. It was because of the last that he was able to indulge in aircraft archeology. The National Aviation Museum is fortunate today to hold some of the aeronautical treasures that Graham found.

During the Second World War, Graham used his mapping skills to plan airfields across the country for the British Commonwealth Air Training Plan, and for his efforts in this, he was named an Officer of the British Empire. In 1946, he became the Canadian technical representative to the International Civil Aviation Organization and chairman of their navigation committee. From 1951 until his retirement in 1963, he was the ICAO adviser to Third World countries — in Latin America, Haiti, Rwanda, and Ethiopia. The Emperor Hailie Selassie awarded him the Star of Menelik for organizing the civil aviation department in Ethiopia.

In 1973, for his achievements in early Canadian aviation and aerial photography, Stuart Graham was named a member of Canada's Aviation Hall of Fame.

---

1 The hull of this HS2L was later recovered by an expedition led by the National Aviation Museum's curator R.W. Bradford and may be seen today on display in that institution.

# GRAY, ROBERT HAMPTON, vc

T he last Canadian airman to be honoured with a Victoria Cross in the Second World War was Robert Gray. It was a posthumous award, made all the more poignant because it took place in the very last week of the conflict.

Robert Hampton Gray was born on November 2, 1917, in Trail, British Columbia. He was educated at Nelson, B.C., and graduated from university in Edmonton, Alberta, in 1940. He enlisted as an ordinary seaman in the Royal Canadian Volunteer Reserve a month after his graduation. That December, Gray began pilot training at HMS *St. Vincent,* Gosport, in England, and two years later he joined his first squadron, No. 757, at HMS *Kestrel.*

In May 1942, he was transferred to 789 Squadron HMS *Afrikaner* and flew with them on postings around South Africa and Kenya. As the only Canadian in a British squadron, Gray came in for a certain amount of teasing about his 'colonial' background and the 'hick town' of Trail that he came from. His boyish appearance and plump figure earned him the nickname 'Hammy', and the Canadian was one of the most popular figures in the squadron.

In May 1944 after leave in Canada, Gray returned to England to be post-ed to the aircraft carrier HMS *Formidable* as second-in-command of 1841 Squadron.

This armoured-deck aircraft carrier was host to four Fleet Air Arm squadrons: 1841 and 1842 flying the American-built, gull-winged Corsairs; and 826 and 828 equipped with Fairey Barracuda torpedo bombers. Almost as he came aboard, *Formidable's* aircraft flew several strikes against the German battleship *Tirpitz* then sheltering in a Norwegian fjord. While the Barracudas launched their torpedoes, Hammy Gray led his crews down almost to sea level, to strafe the German flak defences. He was mentioned in naval despatches for flying daringly low-level strikes against enemy destroyers pro-tecting the battleship. His gun camera recorded that he had almost 'flown up the German's gun barrels.'

To commemorate Robert Gray, the Canadian Warplane Heritage flew this Vought Corsair in British Pacific Fleet colours. (NAM 13771)

In 1945 *Formidable* joined the British Pacific Fleet in operations against Japan. Her armoured deck being able to absorb the kamikaze attacks, the carrier's aircraft ranged far over the Inland Sea area, searching out the enemy. Through most of July, Gray led anti-shipping strikes and strafing attacks against Japanese airfields. On July 28, he scored a direct hit on a destroyer and for this and his exemplary record against the *Tirpitz* he was awarded the Distinguished Service Cross. Then the Fleet withdrew from operations for a week because of typhoons and the dropping of the atom bomb on Hiroshima on August 6.

Gray's squadron had been tasked with maintaining pressure against the dwindling Japanese naval and air forces, but because of the possibility of surrender, were told not to take unnecessary risks. On August 9, 1945, three fighter-bomber sorties were launched against Japanese airfields to prevent more kamikaze attacks from taking place.

At 8 a.m. Hammy Gray led a sortie of eight Corsairs to strafe enemy airfields on the coast of Honshu. They climbed to 10,000 feet and formed into two sections of four aircraft each. Gray was not flying his usual Corsair 119, which was being repaired but he led the quartets in 115. Each Corsair painted dark blue with the British Pacific Fleet roundels, was armed with its usual four .50 machine guns and carried a pair of 500-pound bombs.

While waiting for the command to turn into the wind and fly off the aircraft, a signal was received that provided an alternative target. A reconnaissance aircraft had reported a small convoy in Onagawa Wan Bay, and Gray was notified.

After flying about 150 miles, they made Onagawa Wan Bay, and the pilots spotted a pair of Japanese destroyers with their escorts and several other vessels in the anchorage. They pressed on to the intended airfield only to find that it had already been raided and was a mass of smoke and flame. Gray remembered the destroyers and radioed the others to follow him. His strategy was to swoop down from their present height of 10,000 feet over the anchorage, flash over the hills, and drop their bombs before the destroyers' guns could respond. They would then make their escape to the open sea almost at sea level, back to the carrier.

As each Corsair dived down to the ships, blossoms of flak came up to greet them. The naval guns were joined by the anti-aircraft batteries in the hills overlooking the harbour. Gray led the attack, dropped his bombs on the ship and sped out of the smoke at less than 40 feet above the waves. The whole attack had taken 20 seconds.

Sub-Lieutenant John Blade had just released his bombs and emerged from the chaos when he saw Gray's aircraft on his starboard burst into flames from its port wing. He watched as the stricken Corsair jerked into a starboard bank and then, with its wings ablaze, dive at full speed into the sea. It must have continued beneath the waves at full power as there was no trace of oil on the surface.

The seven numb survivors reformed and flew over the harbour again. They saw that the destroyer that Gray had hit was sinking, and they concentrated on the other vessels at anchor. On return to the carrier, John Blade discovered that his hydraulics had been shot away, and he belly-landed on the deck. Gray had been the only casualty. Five days later the war ended.

In light of his leadership and personal example in pressing home the attack at the cost of his own life, Robert Hampton Gray was awarded a posthumous Victoria Cross on November 12, 1945.

Through the years, in co-operation with the Canadian Embassy in Tokyo and the Japanese government, attempts have been made to find Hammy Gray's Corsair but without much success. To commemorate his bravery, the Canadian Warplane Heritage flies a Corsair in dark blue overall, with British Pacific Fleet roundels and '115' painted in white.

# HARTMAN, PAUL ALBERT

Paul Albert Hartman was born at Grafton, Massachusetts, on November 25, 1918. He was educated at South Portland, Maine, where he also learned to fly. In 1941, like many other Americans, he enlisted in the Royal Canadian Air Force and graduated as a pilot. After a navigational reconnaissance course at Charlottetown, he was ordered to the Royal Air Force Ferry Command at Dorval, Quebec, to fly an aircraft to Scotland.

Hartman completed his operational training in Northern Ireland in 1942 and was then posted to the RAF's 69 Squadron at Malta. The Squadron was flying Wellington bombers in an anti-shipping role, to stop the Germans from reinforcing Rommel's armies in North Africa. The Canadian pilot was awarded the Distinguished Flying Cross in late 1942 for torpedoing an enemy supply vessel through a hail of gunfire from its protecting destroyers. In 1943, Hartman, now a flying officer, returned to Canada to serve as an instructor with No. 6 Operational Training Unit in the RCAF. He later took command of No. 3 Training Squadron and Glider Training Attachment at Cassidy, British Columbia.

At the end of the war, Hartman was named test pilot and attached to the RCAF's Test and Development Establishment at Rockcliffe, Ontario. This was a time when strange and wonderful devices were being developed in aviation — like the helicopter, the flying wing, and the jet fighter. In competition with the United States, Great Britain was testing jet interceptors to meet the new Soviet threat. As a stop-gap measure, the RCAF armed itself with tiny de Havilland Vampires, and prudently sent their test pilots to Farnborough, England, for extensive training in jet aircraft.

Throughout the war, to its credit, the RCAF had kept abreast of the development of jet aircraft, and in 1945 it released specifications for an all-Canadian jet interceptor suited to cold weather. The threat, the airforce and public knew was now going to come from over the polar ice-cap. The Canadian government was forced to the embarrassing conclusion that it could do little with Vampires and piston-engined Mustangs should Soviet MIG-15 and IL-28 jets actually intrude into its airspace.

Paul Hartman in the National Aviation Museum's Avro 504, July 1972.
(NAM 24046)

In October 1950, the RCAF was given the first of the CF-100 Canucks to test. Squadron Leader Paul Hartman had graduated from the Empire test pilot's school at Farnborough and was comfortable with flying the twin-engined jet. As there are in any new aircraft, several defects became evident in the early stages. The teething troubles of the Orenda engine and the buckling of the skin when the wing was stressed were two problems that had to be solved before the CF-100 reached operational status. After a gruelling test program, Hartman flew the first CF-100 to No. 3 OTU at North Bay on July 22, 1952.

Hartman was also involved with the RCAF's other jet fighter, the Canadair Sabre. Designed by the North American Aircraft Company the F-86 Sabre was to the air war over Korea what the Spitfire had been in the Battle of Britain — the classic fighter of its day. Now made by Canadair in Montreal, the RCAF maintained a group of pilots to test fly each new or overhauled Sabre.

Hartman recalls that he first flew a Sabre at Cartierville on March 8, 1951, and found that of all the aircraft he had ever flown, the Sabre was the

To commemorate the 50th anniversary of flight in Canada, on February 21, 1959, Paul Hartman flew a replica of the *Silver Dart* at Baddeck, Nova Scotia. J. A. D. McCurdy met him after the flight. (NAM 18570)

most pleasant, the most reliable, to fly. He rated the Canadair Sabres as better than the U.S.-built ones, and recalled hearing that there was considerable competition among USAF pilots in Korea to fly a Canadian Sabre. Equipping the RCAF squadrons in Europe with the shark-snouted aircraft meant that they were that theatre's 'top guns' in the early sixties. The Sabres consistently outperformed anything that either the rest of NATO or the Soviets had. For the cocky RCAF Sabre pilots, these were the golden years.

In 1952 Hartman was promoted to wing commander and assigned to RCAF Headquarters in Ottawa. The Royal Aeronautical Society recognized his achievements as a test pilot and made him an Associate Fellow. In 1961 he was named commanding oficer of the Central Experimental and Proving Establishment (CEPE). This was the organization that "accepted" the Sabres from Canadair for the RCAF. Hartman had the aircraft flown to the limits, and everything tested. Planes were flown and then re-flown until the CEPE was satisfied that they were up to operational standard.

In 1959 he became famous for piloting a slightly less modern aircraft — a replica of the *Silver Dart*. RCAF personnel had built a non-flying replica at Trenton to commemorate the 50th anniversary flight by J.A.D. McCurdy at Baddeck, Nova Scotia, in 1909. So impressed was the Chief of Air Staff A/M Roy by the replica that he encouraged the builders to make it flyable. A 65 hp Continental engine was fitted, and the *Silver Dart II* was prepared for flight. To Paul Hartman went the honour of piloting the first aircraft ever flown in the British Empire, half a century ago.

The flight took place at Mount View, Ontario, on January 29, 1959. In a tribute to its original designers 50 years before, Hartman remembered that he found it easy to keep the light aircraft straight on the runway. When the airspeed indicator reached 30 mph, the *Silver Dart II* lifted off. He flew it to a height of 6 feet 8 inches for about 150 to 200 feet.

A later flight was not as successful: he lost control of the *Silver Dart II* crashing it but causing little damage. The test flights completed, the little biplane was carried by a C-119 "Flying Boxcar" to Sydney, Nova Scotia, where it was trucked to Bras d'Or Lake at Baddeck. On February 21, Hartman made the ninth and best of his flights on the frozen lake, soaring over half a mile at 43 mph. When he landed, J.A.D. McCurdy was the first to greet him.

On February 23, the actual anniversary of the flight, Hartman took off into a windy sky at 11:30. He described in his own words what then happened:

> Immediately, the wind increased and began to gust to
> 15-20 miles an hour. Caught in a sudden gust, the aircraft

was tossed about 100 feet above the ice before control was partially regained. Full throttle was required to keep the aircraft moving forward against the wind which appeared to be gusting ... The aircraft was almost out of control; the ailerons and rudder were completely ineffective ... the flight was analogous to trying to control a cardboard box in the high wind. A landing was eventually made by closing the throttle and allowing the aircraft to stall onto the ice ... The test and demonstration flying was a most interesting and informative experience, but one which I had no desire to repeat [1]

No longer a test pilot, in 1966 Paul Hartman became commanding officer of No. 115 Air Transport Unit then attached to the United Nations Emergency Force at El Arish, Egypt. Following a stint as base operations officer at CFB Uplands, Ontario, he retired from the air force in 1968.

However, his flying days were not finished. Canadair had just built the CL-215 waterbomber and needed a test pilot. Hartman test-flew it, enjoying its easy handling and ruggedness at low altitudes. He even got to fly a Sabre once more as it was being used as a chase plane for the CL-41 Tutor program. Later, having flown over 120 different aircraft, from the First World War trainers to the latest jets, Paul Hartman opened his own aeronautical consulting firm. Few would argue that no one knew more about aircraft.

---

1 Milberry, Larry, *Sixty Years: The RCAF and CF Air Command 1924-1984* ( Toronto: CANAV Books, 1984), p. 362.

David Hornell (on the right). At age 33, he would have been deferred
from uniformed service in the RCAF. (PL 30823)

# HORNELL, DAVID ERNEST, VC

The use of land-based aircraft in marine warfare was one of the least recognized aspects of the Second World War. It involved endless patrols in slow, overburdened aircraft, struggling against the prevailing winds. There was little hope of survival if shot down into the pitiless ocean below. Unlike other campaigns, the Battle of the Atlantic raged from the first to the last day of the war as the German submarine fleet strove to starve Britain into surrender. Canadian aircrews began reaching RAF Coastal Command as early as 1940 in a variety of aircraft: Hudsons, Hampdens, Wellingtons, Sunderlands and Catalinas. It was in the last that David Hornell earned his Victoria Cross.

Born on January 26, 1910, at Mimico, Ontario, David Ernest Hornell was educated at Mimico High School and Western Technical School in Toronto. He is remembered as an athlete who played rugby for his school and tennis for the Lakeshore Tennis Club. Hornell's particular forte was in track and field, and as a well-rounded student and athlete he was offered a university scholarship. The scarcity of jobs during the Depression might have convinced him to seek a career instead, and he accepted employment at the Goodyear Tire and Rubber Company in Toronto. Had not the urge to enlist in the war proved too powerful to resist, he would have settled into the stable, if humdrum, job for the remainder of his life. At the age of 31, Hornell would have been deferred from uniformed service. Working for Goodyear's he was also in a vital industry.

Neither reason was enough to prevent him from enlisting in the RCAF on January 8, 1941. He received his 'wings' at Brantford, Ontario, that July and was posted to 120 Squadron in Nova Scotia in December. For the following two years he flew Supermarine Stranraer flying boats on coastal patrols along Canada's east coast.

From the beginning of the hostilities, the RCAF's Eastern Air Command was concerned with convoy escort and anti-submarine patrols — as far as their obsolete aircraft could range. German submarines had made forays down the Gulf of the St. Lawrence, and close inshore as at Conception Bay, Newfoundland. But these were more of a nuisance value. The U-boats usually lay in wait for the Britain-bound convoys in the area south of Greenland known as "the black hole." Here the pickings were easier as the U-boats were out of range of even the RCAF Catalinas and, most definitely, the antiquated Stranraers.

Designed in 1935 by R.J. Mitchell, the same man who designed the lovely Spitfire, the Supermarine Stranraer was a metal, twin-engined biplane flying boat. Nick-named 'Stranny,' it was the rugged, slow-moving aircraft that the RCAF needed to patrol both its coastlines. By 1940, because it lacked the range and armament, both the RAF and RCAF had withdrawn their Stranraers from front-line service. To Hornell, as to all who flew her, the Stranraer was reliable, docile, and more seaworthy than the Catalina. Derived from the graceful Imperial Airways flying boats of the 1920s, the Stranraer was, by the Second World War, past its prime and unable to carry the heavy armament necessary in an increasingly scientific war.

On September 22, Hornell left his Stranraer to be posted to 162 Squadron RCAF, equipped with Canadian-built Catalinas, called Cansos. That year he was married to Genevieve Noecker, a music teacher. And for a short while he was 'loaned' to the Boeing Aircraft Company as a test pilot.

On the last day of 1943, Hornell's squadron left for their new home at Reykjavik, Iceland. While an improvement on the Stranraer, the Canso was not suited for 162's mission. Unlike the Sunderland flying boat, it was twin-engined and painfully slow in performance. Its speed was further impeded when the wing-racked radar, Leigh Lights, and bomb loads were added on. Against the U-boat's batteries of 20 mm and 37 mm cannons, all it had were two manually-operated .303 Browning machine guns in the nose turret and one each in the side fuselage blisters.

But what the Iceland-based Cansos did have was the range to bridge 'the black hole.' They were legendary at staying aloft almost indefinitely. In 1942, an RCAF Canso from 240 Squadron in the Shetlands tasked with finding out if there was a German garrison at Spitsbergen, Norway, was in the air for 25 long hours, having covered 2,500 miles.

Starting first as co-pilot, Hornell was given command of his own Canso on May 1, 1944. In anticipation of increased U-boat activity because of the imminent invasion of Europe, three Cansos from 162 were detached to Wick in northern Scotland, to be rotated on a regular basis with other unit crews.

On June 24, Hornell readied for their third sortie in the detachment, flying Canso 9754, coded P. The seven-man crew were all operationally experienced and well integrated into a team. They left Wick at 9:30 a.m. for their

11 hour patrol.

Ten hours later, on the return leg home, the starboard gunner spotted a surfaced German submarine, about five miles distant. Hornell sounded 'Action Stations' on the Klaxon, and the crew manned their weapons as the big flying boat went into an attack pattern. Four miles from the surfaced U-boat, shells began bursting on all sides of the Canso. The German guns had the range to hit the slow-moving target, while the aircraft's .303 Brownings would have been ineffective for another three miles. The U-boat's first hit destroyed all radio equipment, effectively ending all calls for help to Wick.

Shell after shell punished the flying boat, and parts of the starboard wing were torn off. The closer it got, the more accurate and withering was the aim of the German gunners. Then one burst shattered the starboard engine and a second exploded within the fuselage itself. Both Hornell and the co-pilot used all their strength and courage to keep the Canso on course, and bring it close enough to the U-boat to bomb. After what must have seemed like an eternity, the aircraft was within the 1,200-yard range to use its own ineffective guns. Soon the submarine conning tower was sprayed with bullets. But accurate fire for the Canadian gunners was impossible as the Canso shuddered constantly, taking hit after hit.

Then the Germans suddenly ceased firing as the submarine commander swung his ship to port, broadside onto the Canso's course. This was a standard tactic to reduce the chance of being hit by the aircraft's bombs. Hornell skimmed over the deck at about 50 feet releasing the depth charges in a strad-dle. The U-boat was lifted completely out of the water by the explosion, and fell back in a flower of spray.

There was no time to exalt over their victory, for by now the Canadians had other preoccupations. The Canso's starboard engine was aflame and fell completely out of its mounting, the gap filling with gushing oil and petrol that only increased the wing fire. Hornell desperately tried to gain some height but at 250 feet he had to edge the Canso into the prevailing wind to ditch.

The aircraft bounced onto the 12-foot waves, trailing fire and smoke from the starboard wing. Finally, Canso 9754 plunged into the waves and halted. Hornell and the rest of the crew tore out of the sinking plane through the port blister opening. Both dinghies were launched with the emergency rations, the aircraft slid under the waves 10 minutes later.

Then began their ordeal by cold. One of the dinghies blew up and all eight men were forced to cling to the remaining one for survival. Hornell organized his men in shifts: four sat in the dinghy, baling it out while the other four clung to its sides, immersed up to their necks in the freezing North

Atlantic. After an hour of this, it was decided to squeeze everyone into the dinghy and take their chances.

Throughout the night, 60-knot winds whipped up the ocean into 40-foot swells, and the little dinghy was tossed from crest to trough. Just before midnight, a Catalina from 333 Norwegian Squadron sighted them — and some distance away, the survivors of the sunk U-boat. Aided by the firing of flares from the Canadians, the Norwegians radioed for help and dropped a cluster of dinghies over both crews, but by now the pathetic condition of the survivors, both Canadian and German, prevented them from getting out of their own. Then they flashed down with an Aldis lamp the message: "Courage - HSL (High Speed Launch) on way - help coming." and later "U-boat killed."

For the next 14 hours, the 333 Squadron Catalina continued to circle the dinghy watching it helplessly capsize, its crew gradually succumbing to the freezing water.

Air/sea rescues were commonplace in the Battle of the Atlantic, but rarely were actual pickups on the high seas attempted. The Canso, unlike the rugged Stranraer could break its back easily in trying to land in a heavy swell. The North Atlantic was never calm, and the chances of a successful landing and takeoff were very poor. Coastal Command had learned this tragic lesson the hard way when in 1943 a Royal Australian Air Force crew had ditched in the Bay of Biscay. Two aircraft crashed and 13 men were killed by the time the five original crew could be rescued.

Rescue, in whatever form it came, would find a depleted Canso crew. The first to die from the cold was the flight engineer, and after attempting resuscitation, the crew put his body over the side. Some 16 hours after they had crashed, a Warwick rescue aircraft flew over and dropped a lifeboat by parachute. Unfortunately, the parachute's release gear malfunctioned and the boat fell 500 yards away from Hornell's crew. By then, Hornell had lost his eyesight and was succumbing to the effects of prolonged exposure. When another crew member died after many more hours of immersion in the water, Hornell intensified his efforts to save the remainder of his men. Completely blind now and almost unconscious, he strove to comfort and encourage the younger men.

It was then that the exhausted men saw an angel above them. It appeared in the guise of a Sunderland flying boat leading a rescue launch towards them. They had been 20 hours, 35 minutes in the North Atlantic.

David Hornell lapsed into unconsciousness and despite all attempts to revive him by the launch's crew, died 20 minutes after rescue. He was buried in a lonely cemetery near the grounds of Lerwick Hospital, Scotland. Eight days later, the surviving members of his crew having recovered sufficiently in the same hospital, returned to Canada and told of their skipper's courage.

For pressing home an attack with a crippled aircraft, followed by 20 hours of keeping his crew alive in a dinghy at the cost of his own life, David Ernest Hornell was posthumously awarded the Victoria Cross. He was the first member of the RCAF to receive this honour.

On December 12, 1944, his widow, Genevieve Hornell, accepted her husband's bronze cross at Government House in Ottawa from the Earl of Athlone, governor general of Canada.

# MacGILL, ELSIE

As early as 1928, there were many Canadian women who had private pilot's licences. During the Second World War, many more earned commercial pilot's licences, to ferry aircraft to dispersal points across the country. Although Elsie MacGill never piloted an aircraft in her life, her contribution to the development of aviation in Canada far exceeds all others of her sex.

MacGill was born in Vancouver on March 27, 1906, the daughter of the first female judge in British Columbia. She was quite clearly her mother's daughter: she became the first woman to graduate in electrical engineering from the University of Toronto, in 1927. After obtaining her B.A.Sc. she went on to do graduate studies at the University of Michigan.

Elsie MacGill and John Soulsby, the Canadian Car plant superintendent, with the Maple Leaf Trainer, Fort William, November 1939.

Her courses were tragically interrupted by an acute attack of myelitis, and she was confined to a wheelchair. However, this did not deter her and once she had recovered sufficiently to use a cane, she enrolled in the Massachusetts Institute of Technology to study aeronautics.

Highly qualified for her day, MacGill accepted a position with Fairchild Aircraft in Montreal to do stress analysis on the prototype of their first all-metal aircraft. She co-operated with the National Research Council in Ottawa where the aircraft model was tested in its wind tunnel. When the prototype was put through its flying trials at Montreal, Elsie flew with the pilot on all test flights.

In 1938, she was elected to corporate membership in the Engineering Institute in Canada, and in the same year left Fairchild to become the chief aeronautical engineer at Canadian Car & Foundry. She set to work immediately, obtaining a Certificate of Airworthiness of the company's project — the Gregor FDB-1.

Canadian Car had profited from its experience assembling the Grumman Goblin biplane. It had exported the Goblin overseas and even sold several to the RCAF. Building on this in 1938, the company's designer, Michael Gregor, conceived an indigenous fighter aircraft, the FDB-1. Of all-metal construction with flush riveting, the Gregor had a retractable undercarriage. The pilot sat in a shatterproof glass canopy and enjoyed excellent vision because of the gull wing design.

Had circumstances been different, the FDB-1 would have been a match for the monoplanes of the day like the early Hawker Hurricane and Curtiss P-36. Unfortunately, the era of the biplane had ended and there was little sales potential. The RCAF did evaluate it but knew in the coming war it would never rival the German Bf. 109. The ongoing Spanish Civil War also showed that biplanes could not compete against the Messerschmitts. Canadian Car cancelled the project and the Gregor prototype was accidentally destroyed by fire, but MacGill's contribution to the whole scheme had not gone unnoticed.

With the war looming, Canadian Car turned to a more mundane aircraft project — that of building a trainer. Miss MacGill's job was to design a very basic, very sturdy training aircraft that would complement de Havilland's Tiger Moth, then the primary trainer for the burgeoning RCAF.

The result was her Maple Leaf Trainer. This was a biplane with clean lines, very forgiving to fly, and with some relationship to the FDB-1. Built and test-flown at the plant in Fort William, Ontario, the aircraft received it's Certificate of Airworthiness within eight months of its conception, a record at that time. Elsie flew as an observer on all flights and the little trainer is remembered as her creation.

During the Second World War, Canadian Car was awarded the contract to build Hawker Hurricanes. The famous fighter had proved itself sufficiently in the Battle of Britain to warrant a winterized version. Obsolete now, when compared with the new generation of fighters, the Hurricane was flown in Canadian home defence squadrons and also supplied to the Soviet Union. This was the first monoplane that the company had ever worked on, and because of it's winter conversion, would have to be refitted with skis and de-icers. Completely under MacGill's supervision, the program turned out Hurricanes in a record time.

With growing confidence, Canadian Car began bidding for other contracts as well. The failure of the FDB-1 showed that indigenous designs required extensive and expensive research and development, coupled with the marketing that Canadian companies could not afford. In a time-honoured tradition, the Fort William company turned to Curtiss-Wright of the United States to licence-build their Helldivers. Again, MacGill was deeply involved in the retooling of the plant, and in testing of the production models.

In 1941 she was awarded the Gzowski Medal by the Engineering Institute of Canada for her paper entitled "Factors Affecting the Mass Production of Aircraft."

Two years later she left Canadian Car to set up her own aviation engineering consulting business in Toronto. In 1943, Elsie MacGill married another aircraft designer, E.J. Soulsby, a former Canadian Car colleague, then employed by Victory Aircraft at Malton.

In 1946, keeping her maiden name, MacGill became the first woman to be appointed as technical advisor to the International Civil Aviation Association (ICAO). She attended United Nations sessions on airworthiness and helped draft the regulations for the design and production of aircraft. She was also chosen to be the chairperson of the UN Stress Analysis Committee. At a time when airlines were expanding across the world, and jet aircraft were about to enter the scene, this was a crucial job and to chair it was a unique honour — not only for Canada, but for a woman.

In March 1953, the American Society of Women Engineers, in recognition of MacGill's lifelong contribution to aeronautical engineering, made her a honourary engineer and named her "Woman Engineer of the Year."

She remained active throughout her life, not only in the field of aviation but also in the advancement of females in hitherto male dominated professions. In 1967, she was nominated by Prime Minister Lester B. Pearson to the Royal Commission on the Status of Women. Elsie MacGill died in Massachusetts, on November 4, 1980, laden with honours from governments, universities, and industries.

Elected to Canada's Hall of Fame in 1973, her citation read: "Her resolve led her to the top of her profession." It should be added that in a profession dominated by males, she did not let her sex, much less her physical disabilities, stop her.

Donald MacLaren in his Camel. In six short months in 1918, he became
the third-ranking Canadian fighter ace of the First World War. (NAM 8422)

# MacLAREN, DONALD RODERICK

I n the spring of 1918, there was a lull in air activities over the Western Front. The Germans were building up for their final great effort to win the war — the Ludendorff Offensive. All of the established Canadian air aces had been assigned to other duties. W.G. Barker was serving in Italy, W.A. Bishop in England, and R. Collishaw was busy with the administration of his naval squadrons. It was then that the last Canadian air ace of World War I began his career.

Donald Roderick MacLaren was born in Ottawa, on May 28, 1893. The family moved to Calgary, Alberta, in 1898 where MacLaren was educated. After graduating from McGill University, he joined the family fur-trading business at Keg River Prairie, Alberta.

MacLaren was a late entrant into the carnage of the First World War — it was November 1917 by the time he learned to fly, and then he went to France. After extensive training, the 25 year old joined 46 Squadron as a second lieutenant. The squadron was equipped with Sopwith Camels, difficult aircraft to fly, yet deadly in the hands of conscientious pilots.

MacLaren was a phenomenum. Others had survived the Fokker scourge by being good pilots and expert marksmen. They had honed their tactical skills well and gained the self-confidence that only combat experience could give. In a war when an average fighter pilot could expect to last six weeks, MacLaren, at 25 was an old man. Yet without any operational flying behind him, he went on to become the third ranking Canadian ace of the war.

On March 6, 1918, he began his meteoric rise by shooting down a German scout aircraft west of Douai. Four days later he sent an Albatross down in flames. It was the start of a string of victories that seven months later totalled 48 kills and won him several decorations. At a critical time in the development of aerial warfare, MacLaren was promoted to major and given command of his squadron.

By the last year of the war, air fighting was no longer a clash between two knights of the air but rather a confused melee into which whole squadrons were sucked. After three years of butchery, even the opposing air forces were like punch-drunk boxers, hammering each other into oblivion. The day of the wily hunter had ended. Thanks to the mass production of both pilots and aircraft, the RFC was able to operate over enemy territory in swarms. The strategy called for three squadrons to fly together in a triangular relationship. For example, in a typical swarm, MacLaren's Sopwith Camels flew at 15,000 feet in the front and centre, with SE5a's at 16,000 feet to the rear and Bristol Fighters at 18,000 feet behind the SE5a's and to the right of the Camels.

The massive sweeps were met by von Richthofen's Jagdgeschwaders in equally large numbers and authentic air battles took place. Soon the skies over the offensive were criss-crossed by white ribbons of tracer bullets, and black feathers of burning aircraft plunging to the ground.

The duties of being a leader that could shepherd neophyte pilots into the fray were now more important than the skills of individualism and marksmanship. Through that hard fighting summer, as the desperate Germans sought one last opportunity for victory, MacLaren's men found that he excelled at this. By October his record of victories was exceeded only by Barker and Bishop, both of whom had been flying operationally long before he had joined the RFC.

At the end of the War, MacLaren joined the newly formed Canadian Air Force in Shoreham, England. The Royal Air Force was reluctant to see the Canadians branch off into their own air arm , but with Billy Bishop's prestige and persistence, Prime Minister Sir Robert Borden unexpectedly gave his consent and the nucleus of a Canadian Air Force was begun.

Fighter Squadrons 1 and 2 were drawn up, and MacLaren was made flight commander of the first. As with any new organization, the CAF suffered from teething difficulties, especially in terms of a lack of an administrative infrastructure and morale. Much of this was of Ottawa's making.

Rather than equip it with aircraft suited to carry out a variety of functions in a future Canadian environment, Ottawa decided to accept the gift of war surplus RAF aircraft. Even with this, Prime Minister Borden was far from satisfied. The British Air Ministry had, until then, underwritten all costs for the CAF base at Shoreham, but remembering Canadian agitation to break away from the Imperial fold, refused to continue to do so. Faced with this expense, Borden rationalized that the need for maintaining an air force in Canada — a land mass far out of the reach of potential enemies — was more than his government could afford. Accordingly, in 1920, the CAF base in Shoreham was gradually closed down, and MacLaren, like hundreds of other pilots, demobilized to Canada.

By 1921, he had moved to Vancouver and organized Pacific Airways to carry out fishery patrols and aerial surveys for the federal and provincial governments. In 1928, when his company merged with Western Canada Airways, MacLaren became superintendent of the Western Canada division. He expanded their operations deep into the Yukon and the sub-Arctic, sometimes flying the Fokker Universals himself. In 1929, MacLaren and H Hollick-Kenyon flew the first airmail service for WCA between Regina, Moose Jaw, Medicine Hat, Lethbridge, and Calgary.

When his company was taken over by Canadian Airways, MacLaren was promoted to assistant general manager for British Columbia. Created by the Winnipeg financier James A. Richardson, Canadian Airways had an abundance of talent and experience. In many ways, the airline's history is that of Canadian bush flying itself. The other managers were veterans like Romeo Vachon, C.H. Dickins, and H. Hollick-Kenyon. In a bid to make Canadian Airways the national airline, Richardson bought out smaller air companies and moved his headquarters from Montreal to Winnipeg, the geographical centre of the country. When in 1937 Ottawa opted to form its own national air carrier, Trans-Canada Air Lines, Richardson never recovered, either physically or financially. He died in 1939 and Canadian Airways was itself bought by the Canadian Pacific Railway to become the basis for Canadian Pacific Airlines.

Donald MacLaren was recruited by Trans-Canada Air Lines as assistant to the vice-president and took part in the early transcontinental flights in the new Lockheed Electras. In 1940, he was made superintendent of stations and five years later executive assistant to the president. When he retired from TCA in 1958, the Electras had been replaced with trans-Atlantic North Stars and Constellations.

During the Second World War, MacLaren had taken an interest in the Air Cadet League and in 1941 formed the first squadron in Winnipeg. He soon became the league's president and Air Canada presented the D.R. MacLaren Trophy to the League in his honour. It is awarded annually to the most proficient air cadet squadron in British Columbia, a fitting tribute to the First World War ace and bush pilot.

W. R. 'Wop' May. He picked up the name 'Wop' from a young cousin. It was the closest she could get to pronouncing Wilfred. ( NAM 2341)

# MAY, WILFRID REID (WOP)

Called "The Snow Eagle" for daring rescues that only Hollywood might invent, 'Wop' May's colourful exploits in the frigid Arctic are the stuff of legends. From tangling with the "Red Baron" in the First World War to saving a whole town from diphtheria, he is a true Canadian hero.

Wilfrid Reid 'Wop' May was born in Carberry, Manitoba, on March 20, 1896. He had acquired the nickname 'Wop' from a young cousin — it was the closest she could get to saying Wilfrid. Joining the Army as soon as he could, in February 1917, May arrived in England as a staff sergeant in the 202nd City of Edmonton infantry battalion. Like many young men caught up in the grisly bloodbath of the trenches, he fell in love with flying and transferred to the Royal Flying Corps.

On his very first flight with 209 Squadron, he shot down an enemy aircraft and returning home with his guns jammed, encountered the dreaded 'Red Baron', Manfred von Richthofen. Fortunately for the future of Canadian aviation, the novice pilot escaped. By the Armistice, May had a very credible score of 12 aircraft shot down. For his courage while strafing enemy troops in the Battle of Amiens in August 1918, when he was shot in the face, May was awarded the Distinguished Flying Cross.

He returned to Canada after the war and began May Airplanes Ltd. of Edmonton. These were the happy-go-lucky days of barnstorming, and May with his brother Court, took full advantage of the public's hunger for such events. Crowds at exhibitions and rodeos in Edmonton, Brandon, Calgary and Saskatoon thrilled at whatever feats the pair in their Curtiss Jenny, could perform.

In the autumn of 1920, May went to New York to ferry back a new Junkers aircraft for Imperial Oil. It was a hazardous journey made more so as winter approached, and it was late January before he reached Edmonton. In 1921, he was granted a commission in the Canadian Air Force and took a refresher course at Camp Borden, Ontario. The death of his brother Court affected him deeply, and for a few years, May barnstormed again, but eventually he took a job as an inspector for the National Cash Register Company.

W. R. 'Wop' May posing in the Avro Avian after his dramatic flight.
Vic Horner is in the front cockpit. ( NAM 4211)

In 1927, when K. Blatchford, the local member of parliament, formed the
Edmonton Flying Club, May was hired as its only flying instructor. Then, on
New Year's Day 1929, there occurred the drama that made 'Wop' May
famous.

In December 1928, the doctor at Fort Vermillion, Alberta, received an
urgent message by dog sled from the Little Red River settlement — some 280
miles away. A fatal epidemic of diphtheria had broken out in the village. The
doctor knew that he had not enough diphtheria anti-toxin to prevent it from
spreading through the whole territory and he wired the authorities in
Edmonton for extra supplies. To be effective, the toxin had to be delivered
immediately: sending it by dogsled would mean delays of up to two months.
The Department of Health in Edmonton asked May and his friend Vic
Horner to fly the medicine up, if possible. With the temperature at -40
degrees F, May and Horner left Blatchford Field January 3, 1929, in their silver Avro Avian. This was their pride and joy, as it had just been delivered from
the factory in England. It still had an open cockpit — and was on wheels!

The 600,000 units of toxin were wrapped in woollen rugs around a char-
coal burner and, with a special tracheotomy set added, carefully loaded on
board. The overburdened little aircraft, designed by Avro for flying clubs in

England, took a long time to get into the air. It was a 600-mile flight over barren terrain where both pilots knew that neither landmarks, nor even the possibility of rescue before the spring thaw, existed. They flew through blizzard-like squalls that so numbed them with cold that controlling the Avian was almost impossible. After three hours, unable to see any further in the winter darkness, they landed on a frozen lake at McLennan Junction and spent the night there.

The next morning, despite the frostbite wounds, through the grey dawn, May and Horner flew on. Their faces were now scarred and bleeding from being whipped by the biting wind. The tiny Avian finally bumped down the runway at Fort Vermillion at 4:30 p.m., just as darkness set in. The anti-toxin was immediately taken by a fast dogteam to Little Red River, and the settlement was saved.

May and Horner had to make the return trip, again facing the subzero weather. Barely able to see out of their open cockpit, they touched down at Blatchford Field on January 6 to a tumultuous welcome from about 5,000 citizens of Edmonton. Their adventure had been picked up by the newspapers around the world and did much to publicize the vital work that bush pilots performed in the North.

In February, the buoyant May and Horner began Commercial Airways with the Avian and a Lockheed Vega, which had an enclosed cabin. Other mercy flights followed. Twice more he carried anti-toxin to northern settlements. From Carcajou to Edmonton, he flew a mentally deranged women, and her baby, and their doctor. In 1930 for these exploits and inaugurating the airmail route over northwest Canada, May was awarded the Trans-Canada Trophy.

Commercial Airways now purchased three new six- seater Bellancas, and made Fort McMurray their headquarters. Although in midwinter the light made it only possible to fly for some two hours a day, May opened the North to regular flying schedules. Those who lived in the sparsely populated Canadian North came to know his Bellancas well. For May never forgot the code of the bush pilot: for people isolated by thousands of miles of trackless forest and muskeg, you were their only lifeline. If you did not get their supplies to them on schedule, they starved. If you dropped off a prospector, a trapper, or a missionary deep in the bush, you returned for him on the appointed day. No one else knew the person's location. If you didn't return, or even if you were late, the unfortunate man had no choice but to take his chances and set out for civilization, on foot. If you could not keep to the schedule, you made sure someone else did.

When the post office gave Commercial Airways the contract to carry mail to Aklavik, it meant that now the Royal Mail could reach this outpost on the

Arctic Ocean from Fort McMurray in two days rather than the three months it took by dogsled. But despite the mail contract, financial problems forced Commercial Airways to be absorbed by Canadian Airways in 1931.

In February 1932, May was engaged in another adventure that would bring him worldwide attention. He took part in the first Arctic manhunt by air. Flying a ski-equipped Bellanca, the bush pilot carried an RCMP constable in the search for Albert Johnson, known as the 'Mad Trapper of Rat River'. Johnson had just murdered a Mountie and fled into the wilderness. May followed Johnson's tracks by air over the frozen Eagle River for several days. The trapper backtracked often, successfully eluding both the aircraft and the relentless posse on foot behind him. Then, on the last flight of the contract, May saw Johnson cornered by the Mounties, and from high above, witnessed the gun battle. At its conclusion, he landed the Bellanca and picked up the body of the 'Mad Trapper' to fly it back to Aklavik.

Other adventures followed. He brought an aircraft down on one ski in an emergency, and on perhaps the strangest of all missions, he was once empowered as a sheriff's bailiff and commanded to seize control of a ship locked in the ice on the Athabaska River. Following the letter of Admiralty law, May had to land the aircraft by the ship and nail the warrant to the mast. To say that the last action was carried out in spite of rather strong protests from the rough crew would be an understatement! In 1935, 'Wop' May was created an Officer of the Order of the British Empire.

During the Second World War, like a lot of other veteran bush pilots, he volunteered to help with the British Commonwealth Air Training Plan. He was appointed district supervisor of Western schools and later organized an Aerial Rescue School to train first-aid parachute crews. After years of surviving crash landings in adverse conditions, this was a project very close to his heart.

In 1951, he transferred to Calgary to become the manager of the Repair Depot there for Canadian Pacific Airlines. On June 21, 1952, while on a hiking vacation in Utah, he suffered a heart attack and died. Always larger than life, 'Wop' May's skills as a pilot were only matched by his flair for the dramatic.

# McCONACHIE, GEORGE
# WILLIAM (GRANT)

One of the few to make the successful transition from bush flying to the boardroom of an international airline was Grant McConachie. Even as a young man, owning only an old Fokker, he envisioned a polar route to Europe — a dream that 30 years later his Canadian Pacific DC-8s would fulfil.

George William 'Grant' McConachie was born in Hamilton, Ontario, on April 24, 1909. He grew up in Edmonton and read everything he could about aircraft. Grant frequented the Edmonton airfield, begging and sometimes getting rides with famous Canadian bush pilots like 'Punch' Dickins and 'Wop' May. The young enthusiast paid for his flying lessons by looking after their aircraft, finally soloing in a DH Moth after only seven hours instruction. By 1931, at the age of 22 he had his commercial pilot's licence, and he set out for China to fly for Chinese National Airways. On his way, McConachie stopped off in Vancouver to visit his Uncle Harry. This was the first of two fateful meetings in the bush pilot's life. Uncle Harry, not wanting to lose his nephew to what was then a country at war with the Japanese, bought him a dilapidated, second-hand Fokker to start his own airline.

McConachie's Independent Airways of Edmonton came into being in August 1932 with his uncle as one of the principal shareholders. It consisted of the Fokker and a de Havilland Puss Moth and not very much else in the way of material assets. The young man began by transporting salted fish from lakes in northern Saskatchewan to a broker in Edmonton. He barnstormed, he advertised, he flew charter flights — all to keep his little company solvent. But when his Fokker crashed, McConachie was forced to file for bankruptcy, for all that remained of Independent Airways was the tiny DH Puss Moth. Then he met Barney Phillips, a mining promoter.

Grant McConachie with his brand new Lockheed, *Yukon Emperor*. (NAM 7186)

Phillips had a story of a lost gold mine — a tune familiar to every bush pilot. He said he had bought a map from an old prospector to a fabulous gold mine and if McConachie took him and his team to it, Independent Airways' financial problems would be solved forever. Sworn to secrecy and expecting the sheriff imminently to impound his Puss Moth, McConachie flew the prospectors with a two-month food supply to the mine. On his return, he was met by the sheriff, who, despite McConachie's pleas, impounded his aircraft.

As he was the only one who knew how to get to the 'lost' mine, Grant could not tell anyone how to resupply Phillips and his crew in the bush. He persuaded a friend with a Junker F-13 to fly him up, but they had to make a forced landing on the way and the friend returned to Vancouver for parts. Soon McConachie was six weeks overdue and didn't know if the prospectors were dead or alive. In desperation he appealed to a sympathetic pilot to fly him in with supplies. They arrived to see Barney Phillips and his team in the last stages of starvation.[1]

In gratitude, with some of the mine's earnings, Phillips bought McConachie two Fokker Universals to set up United Air Transport and service his mining investments at McClair Creek. With Phillips as a partner, UAT was connected to the mining world and soon supported itself by hauling heavy equipment of every kind to mines in the Yukon and northern British Columbia. By the end of 1935, McConachie was confident enough to purchase a cavernous, all-metal Ford Trimotor — the first multi-engine aircraft to operate in the Yukon. This was the same machine that had been flown by Floyd Bennett in 1928 in his ill-fated attempt to rescue the crew of the *Bremen*.

In 1936, the company started a regular passenger and mail run, leaving Edmonton on Saturdays, flying to Peace River, and returning on Mondays. Through Phillip's business contacts and McConachie's vision, UAT prospered, and more aircraft were purchased. McConachie heard that the post office wanted to extend its airmail service from Vancouver to Whitehorse in 1937. He knew that UAT already had sufficient experience flying trappers from Fort Nelson to their lines in the wilderness. The partners bid on the contract for the Yukon and won it. Operating the Trimotor, Grant McConachie made the inaugural mail flight from Edmonton to Whitehorse on July 5, 1937, in 16 hours.

What distinguished an airline from a bush operation was its ability to keep to a schedule — fly by night and in poor weather. To do this, McConachie installed radio facilities on the UAT airway, equipping most of his aircraft with two-way radio equipment. So impressed were the post office by UAT's reliability that the following year they gave the Vancouver-Fort St. John route to them. A mail subsidy meant regular work for a bush airline and was dearly sought after.

That same year UAT merged with its rivals in the area, Ginger Coote Airways, Cariboo Airways, and later, Yukon Southern Air Transport. The Fokker was replaced by more modern Barkley-Grow floatplanes, as McConachie was always on the look out to acquire better and bigger aircraft. Two famous (and well- photographed) accidents were the conflagration of his Fleet Freighter at Chicago airport and, even more spectacularly, when his pride and joy — the Ford Trimotor was smashed into by an RCAF fighter. The incident took place at Vancouver airport on March 2, 1939, when a Hurricane from No. 1 Squadron was attempting to takeoff. Its pilot lost control and careened into the Trimotor.

With the heavier loads and the limited uses of floatplanes in the spring, McConachie began to look at using wheeled aircraft exclusively. It wasn't long before Lockheed Lodestars, hotel accommodation for passengers, and meteorological reports all began to modernize the former bush company.

In 1941, the United States began building the Alaska Highway and pipeline, and Ottawa fearing its influence formulated a plan to amalgamate the small airlines in the British Columbia/Yukon area. The transportation giant, Canadian Pacific Railway, acquired a controlling interest in the airline and Grant McConachie was made assistant to its vice-president. Following the formation of Canadian Pacific Airlines he became responsible for the 10 airlines that they had purchased. In charge of a fleet of 77 aircraft with various types of equipment, based from Montreal to the Yukon, McConachie succeeded in molding them into one smoothly efficient operation.

But it was only after the war that his airline really began to prosper. With war-surplus pilots and aircraft available, rapid expansion of routes and the airline fleet followed. Seventeen DC-3s were acquired, one of which would remain in service with CPA until 1974. Eleven Lockheed Lodestars were bought, still in their USAAF camouflage colours. Then between 1948 and 1951, there were a series of accidents that almost claimed the airline itself: two of McConachie's Cansos crashed off the British Columbia coast; a DC-3 was blown up in the air over St. Joachim, Quebec, by a timebomb placed on board; and another DC-3 crashed at Okanagan Park, B.C.

For McConachie, who had dreamt of overseas expansion for his airline since its inception, the worst was yet to come. On February 2, 1950, a CPA North Star, attempting to land at Tokyo Airport, ran off the end of the runway and ended up in the shallow water of Tokyo Bay. In this instance all the passengers and crew were saved. Sadly, this good fortune did not recur, on July 21, 1951, when a CPA DC-4 with 37 passengers vanished on a flight to the Orient. After an intense search, the wreckage was found buried in a mountain in Alaska between Silk and Yahutat. There were no survivors.

But the persistent former bush pilot pressed on. When Trans-Canada Air Lines declined to fly across the Pacific, he lobbied CPA's directors to seize any routes that the government airline did not want. In what seemed like a peevish move, Ottawa granted CPA permission to fly to the Orient on the condition that it purchase Canadian-made North Star airliners. In April 1949, McConachie borrowed one of these noisy, uncomfortable aircraft from the RCAF and had it flown on a proving flight to Shanghai. As uncomfortable as the North Star was, the CPA crew landed in war-torn Shanghai just as the Communists arrived, and they were lucky to escape with their lives.

In 1948, as a result of a bilateral treaty between the Netherlands and Canada a Canadian company was invited to operate a regular service between Amsterdam and Montreal. Again TCA, satisfied with its Montreal to London monopoly, declined. Ever ambitious, McConachie seized on this and, using North Stars, flew into Amsterdam, making it the centre of his airline's European operations. Although this was a trans-Atlantic flight, his dream to fly the polar route to Europe was closer to becoming a reality.

With his airline based in Vancouver, it was natural that McConachie looked across the Pacific, and in 1949 added Australia, Fiji, Hong Kong, and Tokyo to CPA'S regular service routes. Grant's youthful ambition to see China finally arrived when in 1949 he flew the inaugural CPA flight to Tokyo and Hong Kong. The Korean War proved lucrative for the Canadian airline; it was contracted by the USAAF to transport American military personnel to Tokyo. Using it's DC-4s and North Stars, the company airlifted over 39,000 troops across the Pacific. Then, too, the large number of refugees fleeing from China filled the CPA aircraft at Hong Kong's Kai Tak Airport. McConachie used the profits made from both these groups of passengers to finance the less profitable Australian leg of the operation.

In late 1949, Canadian Pacific became, after BOAC, the second airline to order a jet airliner — the revolutionary de Havilland Comet 1. It was to be based in Sydney, Australia, and operate between there and Hawaii to link up with the North Star flight to Vancouver. On March 1, 1953, the Comet *Empress of Hawaii*, resplendent in CPA livery, took off from England for Australia, its crew hoping to establish a record for speed on the flight. Two days later as it lifted off from Karachi Airport, the Comet crashed and burned at the end of the runway, killing all 11 on board. McConachie would have to wait until 1961 to see his airline enter the jet age.

The backbone of the airline was the reliable, pressurized DC-6B. With their range, in 1957, Canadian Pacific could fly over the North Pole from Vancouver to Amsterdam. Flights from Toronto and Montreal to Lisbon and Madrid followed. In 1958, six Bristol 175 Brittanias, at a cost of $2.9 million each, were purchased and CPA began its South American service to Mexico City, Santiago, Lima, and Buenos Aries. Eight years after the Comet crash, in 1961, DC-8 jet aircraft were added to the fleet.

Grant McConachie was awarded the Trans-Canada Trophy in 1945 in recognition of his efforts at opening up the North to civil aviation. He died of a heart attack at Long Beach, California, on June 29, 1965. The following day, on what was the last flight of a great Canadian aviator, a CPA DC-6B flew his body home. Grant did not live to see the merger between Pacific Western Airlines and Canadian Pacific nor the effects of the recession on his airline. In his lifetime, Canadian Pacific Airlines had expanded from a bush airline to the seventh largest in the world, becoming a serious rival to the government's TCA. It was a far cry from flying fish in an old Fokker. In 1968, the new highway to Vancouver International Airport was named "McConachie Way" in his honour.

---

1 Corley-Smith, *Bush Flying to Blind Flying: British Columbia's Aviation Pioneers* (Victoria: Sono Nis Press, 1993), p. 177.

# McCURDY, JOHN ALEXANDER

Born barely two decades after Confederation, this Nova Scotian is forever linked with the creation of our national air force and aviation industry. Whatever his motives, if ever one man can be considered the Father of Canadian Aviation, it is John McCurdy.

Born at Baddeck, Nova Scotia, on August 2, 1886, McCurdy showed great promise as a student and was sent to Toronto's School of Practical Science. There he met Frederick Walker 'Casey' Baldwin, who was also interested in the science of flight. McCurdy's father was secretary to the distinguished scientist Alexander Graham Bell, and it was he who persuaded Bell to employ the college students. During the summers in Nova Scotia, the young McCurdy and Baldwin were fortunate enough to assist Bell in experiments with his tetrahedral kites.

After graduation as a mechanical engineer in 1907, McCurdy and Baldwin were invited by Bell to form the Aerial Experimental Association of Halifax. Mrs. Bell had come into some money and offered to finance the association. They were joined by two Americans, Glenn Curtiss and Lieutenant Thomas Selfridge.

The association moved to better quarters, at Hammmondsport, New York. A large man-carrying kite *Cygnet* was built and flown. This was followed by their first biplane the *Red Wing*. On May 28, 1908, McCurdy made the first powered flight on the association's second experimental aircraft, the *White Wing*. The five then redesigned the wing structure and built a fourth experimental aircraft, the *Silver Dart*. The distinguishing feature of their latest creation was that it had a water-cooled, eight-cylinder engine. On December 6, the *Silver Dart* was test- flown by McCurdy at Hammondsport.

Then the aircraft was transported to the wintry fastness of Baddeck. On February 23, in a historic moment, John McCurdy took off in the *Silver Dart* from the frozen ice on Bras d'Or Lake. More than 146 people had gathered to watch. This was the first heavier-than-air, powered flight in Canada — and the whole British Empire. Flying three-quarters of a mile at a height of 60

J. A. D. McCurdy, the father of Canadian aviation, demonstrating his *Silver Dart* at Petawawa army camp, June 1909. (NAM 2177)

feet, McCurdy's speed was estimated at 40 mph. By March, McCurdy would operate the *Silver Dart* several times at distances of over 20 miles.

But the association was breaking up. Poor Selfridge, flying with Orville Wright in Washington, would be killed in a crash, the first fatality of the air age. Curtiss, a motorcycle engine manufacturer, returned to his factory at Hammondsport with plans of starting his own aircraft company. As Bell went on to other experiments, the Aerial Association finally dissolved at the end of March. John McCurdy and Casey Baldwin then formed Canada's first aircraft manufacturing company, The Canadian Aerodrome Company of Baddeck, Nova Scotia.

Dr. Bell did give the two young men a boost by speaking at the Canadian Club in Ottawa before the governor general and the minister of finance. It was discussed in the highest circles in Ottawa that Canada should adopt some sort of aviation policy and Canadians be given the opportunity to "demonstrate their aeroplanes on the Government's behalf."[1]

That summer the Baddeck company built its first biplane, *Baddeck 1* and McCurdy planned to take it and the *Silver Dart* for demonstrations in

Ottawa. By mid June the partners transported both aircraft in pieces to Petawawa Army Camp in Ontario. The military, in a rare moment of generosity, wished to show that it was as forward-minded as anyone else and even approved the expenditure of a lavish $5 for lathes and rolls of tarred paper to assist in the rebuilding of what was called 'the aerodromes.'

On July 26, 1909, news came that Louis Bleriot had conquered the English Channel, and patriotic fervour in Canada called for a national feat in this new science. The second day of August dawned clear and bright and the reporters and photographers gathered at Petawawa to witness the *Silver Dart's* historic test flight. Watched by the military and press, McCurdy swung the propeller and climbed on. He first flew half a mile at a height of 10 feet before landing. Then he took Baldwin up as a passenger — the first in Canadian aviation history, at a speed of 50 mph. On his third flight, he took the local foreman up. On the fourth and last flight, the rising sun blinded McCurdy and he hit a grassy knoll, crashing the aircraft onto its starboard wing.

While they were yet to see the second biplane, *Baddeck 1* in action, it seemed that the military had made up their minds. The aircraft's engines would frighten the cavalry's horses, to the detriment of the conduct of a war. The navy was even less enthusiastic, pointing out that the possibility of so fragile a contraption harming an armoured battleship was remote, to say the least.

Nevertheless on August 12, the Company's other aircraft *Baddeck I* was demonstrated before a military audience that remained cool to the latest trial. When McCurdy stalled and crashed, the official party caught the train back to Ottawa with deep misgivings.

The dejected young inventors returned to Nova Scotia, and by November 1909, they were almost penniless. There, far from Ottawa, everything worked well, and with *Baddeck 2* some 50 flights were completed. After various entreaties by McCurdy, Ottawa sent an officer to view these experiments. To prove how reliable the aircraft was, the officer was even carried aloft and proceeded to write a favourable report on the aerial experiments.

However it was in vain. The Cabinet rejected the military's proposal to finance the 'aerodrome.' Now, completely without funds, Baldwin and McCurdy closed down their company in 1910 and Baldwin went to seek his fortune the United States. At a time when England, France, and Germany were actively financing their own air arms and aviation industries, the unique talents of this duo would be lost to Canada forever because of bureaucratic timidity.

McCurdy renewed his association with Glenn Curtiss and, flying his aircraft as one of the earliest barnstormers travelled widely across the United

States. Besides becoming the first Canadian to hold an American pilot's licence, on August 27, 1910, the Nova Scotian made the first successful radio communication between an aircraft in flight and a ground station over Sheepshead Bay, New York. The following year, McCurdy would attempt the first ocean flight — a distance of 96 miles from Key West, Florida, to Havana, Cuba. Using a plane built by his colleague of the Aerial Experimental days, Glenn Curtiss, McCurdy almost made it completely across, but his engine began to lose oil and seize up. He was relieved to be rescued almost immediately as the aircraft had come down in shark-infested waters. Despite this, it was the longest flight over water to that date.

By the outbreak of the First World War, the Canadian Aviation Corps had not only been formed but also shipped its only aircraft — an American Burgess-Dunne — overseas to England.[2] The Canadian minister of militia, Sir Sam Hughes, was overwhelmed by the tasks of organizing the Canadian Expeditionary Force in 1914 and had little time or sympathy for an aviation corps. The brief flirtation with the Burgess-Dunne was about all the effort he was willing to give.

Now firmly associated with the Curtiss Aeroplane Company of Hammondsport, New York, McCurdy knew that the timing in Canada was right. He lobbied Hughes constantly, proposing that a wholly Canadian air army be trained and equipped to fight in the war. Hughes liked the idea of a national air force being set up with a national air industry to support it and passed the proposal on to Prime Minister Sir Robert Borden. McCurdy then lobbied the prime minister's office, writing that both these aims could be conveniently fulfilled if his company was given the contract to set up a Toronto factory to build aircraft and a school to teach Canadians to fly them.

Borden, also preoccupied with more pressing matters, passed this proposal onto the British Admiralty. From McCurdy's viewpoint, this was the best thing that could have happened. The Admiralty was beginning to regard the German zeppelin threat with some alarm, particularly in the balloon's role as a spy over the Fleet. The Royal Naval Air Service not only offered to buy 50 of McCurdy's aircraft but also promised to enlist all the graduates of a Curtiss school.

With McCurdy as managing director, Curtiss Aeroplanes of Toronto opened its doors on April 12, 1915. The Toronto school began its flight training from Hanlan's Point on a harbour island on May 10. Advertising for pilots for the RNAS was inserted in Canadian newspapers. Although the Admiralty limited recruiting to "British subjects of pure European descent between 19 and 23 years of age" there was no shortage of candidates. Using three Curtiss F-type flying boats, training progressed efficiently all summer. But with the winter, flights were impossible, and McCurdy looked to moving the school to Bermuda. When this wasn't feasible, the pupils hung around Toronto impatiently waiting for the spring. The aviation school chose a site at Long Branch just outside Toronto to train pilots on wheeled aircraft. The Curtiss Company built three hangers on an airfield that may be called Canada's first. The wheeled trainers were Curtiss JN3s built at the factory on Dufferin Street.

McCurdy, perhaps thinking that he should be talking to the dog and not the tail, now went to London. Here he lobbied the British to convince the Canadian government to launch a Canadian flying corps. If the truth be known, he hoped to get out of the aviation school business altogether as it had brought him little profit and much bad press. The British were bemused by what they considered a purely 'colonial' problem and gave him short shrift.

Politicians on both sides of the Atlantic suspected that the aviator, for all his patriotism, was first and foremost his own man — or Glenn Curtiss's man. As an advocate for a purely Canadian flying corps, no one stood to gain more financially then he and Curtiss. Morever, McCurdy's incessant lobbying campaign was seen by the staid Edwardian establishment as being not quite gentlemanly. He was accused of misrepresenting to the British his influence with the Canadian government and vice versa.

By 1916 McCurdy's role as a spokesman for a national air arm was tainted with commercialism, and did not garner much sympathy for the formation of a Canadian air force.

It was the short life span of pilots on the Western Front combined with the increasing use of air power that finally convinced the Canadian and British governments into buying out the both the Curtiss air school and factory. The Royal Flying Corps Canada scheme, the most important development in air history in Canada to this point, would train hundreds of pilots on a scale beyond McCurdy's dreams. The Curtiss Long Branch school would be taken over, and the factory was renamed Canadian Aeroplanes Ltd. Work was begun on mass production of the Curtiss JN4D or 'Jenny.' To the barnstormers of the 1920s, it would forever be remembered as the 'Canuck.' McCurdy's lobbying would be overtaken by events. Whatever his motives were at the time, in retrospect it is safe to say that he planted the seed for a Canadian air corps and industry.

After the war, John McCurdy founded an aircraft supply school in Toronto, and in 1928 worked for the Reid Aircraft Company in Montreal. A year later, he merged this with his old employer, Curtiss, to form the Curtiss-Reid Company. He became assistant director general of aircraft production in the Department of Munitions and Supply from 1939 to 1947. After the

Second World War, he was appointed lieutenant governor of Nova Scotia, an office he held until 1952.

John McCurdy was awarded the McKee Trophy on the 50th anniversary of his historic flight in the *Silver Dart*. He had the pleasure of witnessing RCAF test pilot Paul Hartmann fly a replica of his aircraft on a frozen lake at Baddeck just as he had done half a century ago. He died in 1961, having seen the air age in Canada born, and through his efforts, flourish.

---

1 Parkin, H., *Bell & Baldwin* (Toronto: McGraw-Hill Ryerson 1950), p. 274-31.

2 The British, whose aviation technology was far advanced, ignored the Burgess-Dunne and allowed the first Canadian-owned aircraft to sit in a hanger until it rotted away in the damp climate.

# MacINNIS, GERALD LESTER

It has been calculated that the construction of the Distant Early Warning Line has been the largest undertaking of the last 40 years. That it was done in 29 months, under bone-chilling climatic conditions, under the threat of atomic warfare, makes it even more astounding. Its success is due to hundreds of brave men, one of whom was a DC-3 pilot named Jerry MacInnis.

Jerry MacInnis was born on June 2, 1914, in Amherst, Nova Scotia. Between 1936 and 1941, he was the superintendent of a fur farm at Port Meunier on Anticosti Island. When he enlisted in the RCAF in 1941 he was chosen for observer training, but a year later he was commissioned as a pilot officer.

His first posting was to 117 Squadron on the East Coast on anti-submarine duties. Holding both an observer's and pilot's Wings, he flew Cansos over the North Atlantic. The patrols were all the more dangerous because of the imposed wartime radio silence and the lack of navigational aids. In July 1944, Jerry was made a flight lieutenant and seconded to British Overseas Airways Corporation as an instrument flight instructor. BOAC was then operating Liberators across the Atlantic, and after the war strove to regain its long-distance commercial routes using Yorks, Argonauts, and Constellations. Jerry instructed the British crews on these aircraft, but in 1950 he returned to Canada to farm.

Unable to stay away from aviation for long, he joined Maritime Central Airways (MCA) in 1951 to fly their war-surplus DC-3s. With his experience on British and American multi-engined aircraft and his navigational skills, MacInnis was chosen to land the site construction parties that would build the DEW Line radar chain.

The electronic warning "fences" across the Canadian North were the Pine Tree Line, the Mid-Canada Line, and the Distant Early Warning Line. With the start of the Cold War, the United States and Canada co-operated in building a radar network that would give instant warning to the approach of Soviet

Was it ever so simple? Passengers disembarking from a Maritime Central Airways DC-3 that Jerry McInnis would use on the Dew Line. (Nam 5558)

aircraft or missiles across the Polar Ice Cap. By far the most critical of the three chains was the most northern, the DEW Line. It lay along the 69th parallel of north latitude from Alaska across Canada to Baffin Island — about 3,000 miles long. It was made up then of some 60 electronic sites on the average about 50 miles apart. Domed antenna towers rose like a giant's golf balls across the tundra, surrounded by supporting base camps of cabins, generators, and fuel-storage tanks. It was undoubtedly the greatest construction project ever undertaken on this continent. The cost in the 1950s of this huge project was $500 million. Except at a few sites, everything from cement to bulldozers, from medicine to radar valves, had to be carried by air.

When MCA got the contract in 1955, it was to provide transport for 17 of the larger radar stations. The aircraft it had available were DC-3s, Curtiss C-46s, Cansos, DC-4s, and the Avro York. Also involved were the United States Air Force with their cavernous C-124 Globemasters and the RCAF's C-119s.

Flying out of Mont Joli, Quebec, MacInnis used MCA's DC-3s to take advance parties to all 17 locations. Equipped with skis as well as wheels, his strategy was to fly the DC-3 low over the barren terrain to the co-ordinates given. Then he would circle until the engineers decided whether or not to establish a camp there, finally returning with a construction party to land on the site — if he could find it again!

Because of the drifting snow and Arctic winds, locating the original sites usually required two or three flights to erect a radio beacon. The long distances involved and the absence of any aid meant that the pilot's fuel supply was critical, and aircraft always flew with full fuel tanks. Sub-zero temperatures and 80-mile-an-hour winds made the building of anything in the Arctic a grim task.

MacInnis and his DC-3 became well known for shuttling between the railhead and the sites in the worst weather, carrying survey teams and their equipment. He looked for safe landing places, fresh-water sources, gravel for the runways anything that would help facilitate construction of another DEW radar base. The initial landing was especially dangerous for MacInnis: he had no way of knowing how deep or compact the snow was.

Baffin Island, the largest of the Arctic Islands, is the size of a prairie province, some 200,000 square miles. It contains mountains that rise to 10,000 feet, fiords that disappear into the hinterland, glaciers that twist and turn across the rugged topography. Hours of daylight for flying are very short — a maximum of three out of twenty-four in the winter. High sea-ice meant that the only ships that could make it through to supply the sites were icebreakers, and then only in the summer. To erect permanent habitations on it and keep their personnel supplied, by air, tested the courage and ingenuity of all involved. That MacInnis would be given the nickname 'The Arctic Fox' by the engineers he flew is a testament to his exceptional skill.

Establishing a base camp meant that Jerry would drop off three men on the initial trip with as much food and fuel as the aircraft could carry. The pilot then took off to bring back more supplies. As there were no radios, the men left behind knew that their lives depended on MacInnis finding them again and being able to land once more.

A typical example was locating Site 30 (Foxe). On March 23, 1955, MacInnis and his co-pilot, Dave Hoyte, circled over the trackless snow on the location that had been pinpointed on the map. They flew low over some ice and decided that it was safe enough to land. Leaving the engines idling they got out and walked about.

There was a huge deposit of limestone gravel, good for construction. Marking the point with a yellow flag tied to an old broom, MacInnis took off for the base to organize the airlift. He briefed the crews of the other aircraft like this: "We explained that we would take off at dawn the next day at thirty minute intervals, cross to the north shore of Southampton Isle, then follow the anchor ice to Foxe, 300 miles further north, when they would see a yellow flag." After warning them that there were ice hummocks 10 to 20 feet high and that the flag was at the upwind end of the strip, the freighters set off.[1]

Many times heavily laden DEW Line aircraft sank into the snow. A DC-6 ran out of fuel and crash-landed on the ice. Flying tanker aircraft was like playing Russian roulette, realizing that the odds on the volatile gasoline igniting were very high. In a dramatic mishap, a heavily laden Globemaster, while on final approach into Cambridge Bay, landed short and hit a bump. The wings fell off and the two-storey fuselage continued on down the runway, unperturbed! The USAF tried air-dropping heavy equipment to the construction sites, but the nylon shrouds of the parachutes became brittle in the cold weather and snapped off as soon the canopies opened. The first time this occurred, a D-4 Caterpillar tractor fell straight through eight feet of ice.

The stories of MacInnis's exploits among the construction crews grew with every month. Once when co-operating with a party of engineers looking for a landing site, he heard one remark : "I wish I knew the precise height of that peak." At that moment, MacInnis dived onto the site, roared across the terrain and pointing to the altimeter said: " Take 50 feet off that!"

Over 142,000 tons of freight were airlifted to the DEW Line in the 29 months it took to construct it. The electronic chain was operational, on schedule, and integrated into the NORAD system by July 31, 1957. MacInnis had nothing but praise for the Canadian and American personnel who built it. They in turn trusted his judgement to find suitable sites and relied on his ingenuity to locate their advance teams again. To his credit, he never failed them.

The airlift concluded, Jerry MacInnis was transferred to Montreal as MCA began to expand into trans-Atlantic charter flights. In 1959 he became senior air carrier Inspector for the Department of Transport, and in 1974 was promoted to chief of flight operations in Ottawa.

The DEW Line, largely unknown and now forgotten by those it sheltered throughout the Cold War, remains a monument to the men who built it. The intricate web of aircraft and men, both Canadian and American, that built it are well represented by Jerry McInnis and his Maritime Central Airway's DC-3.

---

1 Milberry, Larry, *Aviation in Canada* (Toronto: McGraw-Hill Ryerson, 1979), p. 73.

# McKEE, JAMES DALZELL (DAL)

Forever remembered in Canadian aviation history for the trophy he pre-sented, this American millionaire was a firm friend to the onset of the air age in Canada. The historic trans-continental flight that he and Squadron Leader A.E. Godfrey made in 1926 opened up the country to avia-tion.

James Dalzell McKee was born in Pittsburgh, Pennsylvania, in 1893. Son of a wealthy family of investment bankers, McKee graduated from Princeton and learned to fly in the United States Army Air Corps during the First World War. Handsome, over six feet tall and witty, 'Dal' had a bright future in Pittsburgh social circles. Yet his one ambition was to master the art of flying float and sea-planes.

As a wealthy 33 year old, when he first came to Canada in 1926, it was in his private aircraft. With his Douglas Corps Observation float-plane, he and a companion had planned to fly from Washington to Hudson Bay. But when they landed in Sudbury, the Douglas would not take off from freshwater with such a heavy load. Disappointed, they flew instead to Montreal and left the aircraft at the Vicker's plant. The Douglas Aircraft Co. in Santa Monica advised McKee that if the rear pontoon struts were shortened, take-off perfor-mance would be much improved. Further, they wanted the Douglas returned to their California factory for this experimental work.

McKee thought that he might make this flight in a float-plane across the continent, through Canada, and got in touch with the controller of civil avia-tion in Ottawa. The Department of National Defence was very interested in the possibilities of a float-plane route across the country as this would serve to reinforce air force bases on the Pacific.

They offered maps, refuelling facilities at RCAF bases across the country, and Earl Godfrey, an RCAF officer, to accompany him. Ottawa arranged for Imperial Oil to ship aviation gasoline in advance to refuelling points along the way, and asked the Ontario Provincial Air Service (OPAS) for landing privi-leges at their seaplane facilities.

Dalzell McKee on the right, with Lt. Earl Hoag. (NAM 4969)

Captain McKee and Squadron Leader Godfrey took off on September 11, 1926, at 1500 hrs. EST from the Vickers plant in Montreal. They had hoped to fly as far as Sudbury before evening, but McKee had been delayed by customs at the Canadian border and only arrived at the plant an hour before departure. The Douglas float-plane had been re-fitted with a Liberty engine and was on wooden floats. Because the engine was watercooled, it had to be drained every evening and refilled in the morning to prevent the oil from freezing.

Faced with the loss of time and having to drain the engine, they decided to make for Ottawa the first night, landing at Shirley's Bay at 1630 hrs. The next day they flew 100 miles in the pouring rain along the railway to Lake Traverse. There they tied up at a beach and spent the night with the fire ranger.

On the 13th, with the sky clearing, they flew to Sudbury, landing at the OPAS facilities at Ramsay Lake. The next day with the weather closing in, they came down to 500 feet to follow the railway line from Sudbury to Sioux Lookout. Sometimes unable to take off because of lack of wind or waves, they alighted at lake after lake, always flying over water or the railway.

At Lac du Bonnet, Manitoba, they spent a day at the RCAF station. On the following day they left Prince Albert to arrive at Lake Wabamun, west of Edmonton. Fortunately for posterity, RCAF Flying Officer 'Punch' Dickins, in an DH 4B aircraft, was able to photograph the historic flight on this leg of the route.

On Sunday the 19th, the weather was clear, perfect for navigating through the Rockies. They set off at midday through the magnificent scenery, most of which towered above the tiny aircraft. It was an experience that neither man ever forgot. When they reached Hope, British Columbia, they could see the ocean. On Sunday, September 19, at 1520 hrs. Pacific Time, McKee and Godfrey alighted at the RCAF Station at Jericho Beach. The first float plane flight across Canada had been accomplished!

Ahead, there were troubles with the Liberty engine and water leaking into the battered pontoons. But by September 24 the pair were able to taxi up to the pier at San Francisco's harbour. Because the city was enveloped in fog throughout the next week, McKee decided to end his flight there and had the Douglas company send a pilot to fly the aircraft to Santa Monica. In two weeks they had flown 3,955 miles in 43 hours and 49 minutes.

In 1927, Captain McKee and Earl Godfrey again planned a record-breaking flight through the Canadian wilderness. This time they wanted to fly three aircraft from Montreal to the Yukon, across to Alaska and down to Vancouver before returning to Montreal. The Douglas was re-equipped with a Pratt & Whitney engine and Shorts floats.

The accompanying aircraft were to be two Vickers Vedettes seaplanes from Montreal that both McKee and Godfrey wanted first, to gain experience. With an American, Earl S. Hoag, they planned to use Lac la Pêche near Grand'Mère, Quebec, to practice alighting and taking off.

On June 9, 1927, Godfrey flew one Vedette to Ottawa and landed at Shirley's Bay. He was to wait for McKee to bring the other Vedette there. McKee and Hoag first flew to Lac la Pêche to practice landings. It was evening by the time they got to the tranquil lake. As neither had experience with seaplanes, they misjudged the distance from the water and struck the lake with such an impact that the Vedette broke in two. Hoag was pulled out and treated for shock, but McKee's body wasn't found until the next day.

Earl Godrey flew to Pitsburgh to attend McKee's funeral. Both he and Canada had lost an ally and good friend. The American will always be remembered not only for that historic cross-country flight but also for the Trans-Canada Trophy that he donated. To aviators, it has always been known as the McKee Trophy.

# MYNARSKI, ANDREW CHARLES, VC

Twice in this century, war had put the spur to the development of Canadian aviation both at home and abroad. But in spite of Canada's contribution in the Second World War of men and material, it received little recognition at the councils of the major players in that conflict. Overseas, there was only one Canadian air vice marshal and a single bomber group to publicize the nation's coming of age. For a nation not given to hero-worship, our national identity is linked to the achievements of ordinary airmen — and more so to those who, like Andrew Mynarski, earned the Victoria Cross.

Andrew Charles Mynarski was born in Winnipeg, Manitoba, on October 4, 1916. He worked as a furrier before enlisting in the Royal Winnipeg Rifles in 1940. In 1941, he transferred to the Royal Canadian Air Force and was accepted as a wireless air gunner in a bomber squadron. To understand fully the situation that placed quiet, cheery Andy Mynarski in a flaming Lancaster on June 12, 1944, one must set the scene.

In 1943, Mynarski was assigned to 419 Squadron, part of 6 Group, the all-Canadian bomber force, deeply engaged in the Allied bombing strategy campaign. In a move that would seal the destiny of hundreds of Allied airmen (and thousands of German civilians), on February 22, 1942, the Royal Air Force bombing squadrons that 6 Group would become part of were put under the command of Arthur Travers Harris.

At a time when Prime Minister Winston Churchill was being advised that the war could be lost over control of the Atlantic, Harris singlemindedly stressed that the war could only be won by devastating the enemy's cities. As a First World War pilot, he had seen the massacre of the Somme from above, and he believed that destroying the enemy's ability to wage war on the home-front was more important than defeating armies in the field. Stubbornly he

The *Ruhr Express*, the first Canadian-built Lancaster arrives in Britain. In the VIP party (seventh from the left) is the high commissioner to London, the Hon. Vincent Massey. (PP 010782)

pioneered tactics that used bombers like schools of locusts, overpowering the enemy defences with sheer numbers. In the days when news from the other fronts was dismal, when Stalin was goading the Allies to open a Second Front, Bomber Command was the only way to hit back at the Nazi war machine.

Despite the fact that his 'meatgrinder' tactics would kill many crews it should go on record that the bomber squadrons — Canadian and British — respected their new leader. They saw in him a fellow pilot who wasn't afraid to try the new radar innovations and who was utterly convinced that his way of waging war was the only way. His faith in urban bombing led him to grandstand events before Churchill — and jealous Coastal Command, who felt that the aircraft could better be employed hunting U-boats in the Atlantic. Nothing was too much for Harris's 'boys' to do — the thousand bomber raid over Cologne with a loss rate of 3.9 percent, the Ruhr raid with a loss rate of 3.2 percent or the suicidal U-boat pen raids with a loss of 6 percent.

The roots of how the Canadian 6 Group were thrown into this killing machine, stretch back to May/June 1942 and the Ottawa Air Training Conference. Here, British and Canadian delegates would agree on the formation of more RCAF bomber squadrons. Both Prime Minister Mackenzie King and the Minister for the Air C.G. 'Chubby' Power, with the losses of the First World War in mind, insisted on some control over Canadian operations overseas.

The British, especially Harris, would have preferred to see the BCTAP graduates, now coming on stream, fully integrated into the RAF. A new bomber group hastily formed, would not have the experience or tradition that the RAF had built up over the years. Besides, London felt that creating a new group for the sole purpose of satisfying politicians in far-off Ottawa was sure to become a liability. Viewed from Whitehall with their global problems, the birth of 6 Group seemed like smallmindedness. But to the Canadian public, forming the first non-British force in Bomber Command was a potent symbol of Canada's emerging status as an equal partner with the mother country. Both sides were in their own ways correct, and this was proved by events over Europe in the years to come.

On New Year's Day 1943, the Canadian bomber squadrons in England became part of 6 Group. Equipped with Wellingtons and Halifaxes, the eight (later thirteen) squadrons operated from bases in Yorkshire.

That spring it seemed that Harris would be proved right. In the first months, because of flying accidents, high losses, and poor serviceability only 59 percent of the group's bombers were operational on any given day. 419 (Moose) Squadron had been formed in December 1941 and, by the time Wireless Air Gunner Andrew Mynarski joined it, had already been bloodied on the Ruhr and Berlin raids. With 420 Squadron, 419 operated Halifaxes from Middleton St. George, later switching to the Canadian-built Lancaster Xs. In fact 419 had the honour to operate the *Ruhr Express* the first Canadian-built Lancaster. Steadily, despite severe losses, the efficiency of 6 Group was improved.

The flak, the searchlights, and the nightfighters were taking their toll on the crews, but with the BCATP and Allied aviation industries in full flow, there was always a steady stream of replacement men and machines. This was poignantly illustrated when new Canadian crews arrived at a base and found that there was no accommodation prepared for them. They were told to sleep in the mess that night because, as there was a big raid 'on,' they could take their pick of accommodation the next morning.

After several sorties, the crew of the Lancaster KB726, A-Able, had been welded together as a close-knit, experienced unit. They were seven strangers,

Andrew Mynarski, awarded the Victoria Cross posthumously, served as the upper turret-gunner in a Lancaster of 419 squadron. (PL 38261)

each of whom realized that their own lives depended on the other six. In the impersonality of the massive bombing stream as you watched the deadly fireworks display over some German city, as you saw aircraft all around you get 'coned' by searchlights and shot down, you treasured the friendship within the safety of your crew.

The gunners who were sometimes trained as wireless operators, sat in isolated fear at the extremities of the noisy Lancaster. They wore electrically heated suits and were the eyes of the crew. Firing their machine guns and attracting the attention of patrolling nightfighters, was if possible, best avoided. German fighter pilots, like their counterparts in the Battle of Britain, were guided by the latest radar and intent on killing those who would lay waste to their homeland. The idea was to get in, drop your bombs, and get back intact — fast!

On June 5/6, the invasion of Europe began. Harris was ordered to pull his squadrons off the punishing hinterland raids and destroy enemy transportation centres in France. Although still convinced that area bombing was the most effective way to win the war, he gave in. Every Bomber Command aircraft was now in the air over France as the target list grew: railway marshalling yards, port facilities, coastal batteries, railway bridges...

On June 12/13, 6 Group was ordered to bomb the Arras area. Its 89 aircraft encountered heavy flak and a still-potent Luftwaffe. Nightfighters shot down six of its bombers. The situation in Cambrai was the same.

At 419 Squadron, the Lancaster A-Able, now on its 13th sortie[1] was piloted by Flying Officer Art de Breyne, its crew welded into a efficient camaraderie. The bomber, laden with 9,000 pounds of high explosives got airborne at 9:44 p.m. The weather was fair, and good visibility had been promised over the target. Unknown to him, Andrew Mynarski`s commission as a pilot officer had come through the previous day.

Crossing the French coast, de Breyne began his approach onto the target. The flak defences had sighted them and were providing a heavy barrage of shells at the oncoming bomber stream.

Just as they were to reach Cambrai, a Ju88 attacked from below and cannon shells smashed into the port wing and fuselage. With the explosions, the aircraft began to burn furiously. Soon the port wing was in flames and both engines had stopped. The Number 2 fuel tank between the engines was also on fire. In vain, de Breyne tried to read his instruments but the electrical system was out. He ordered the crew to bail out. Now the intercom wasn't working. The pilot yelled at the flight engineer to pass the word to the others as the bomber began its final spin. He used the emergency warning lights to convey the abandon-ship order, especially to the gunners. With the bombs on board there was little hope of surviving a crash landing. Yet de Breyne held the burning aircraft as steady as he could, allowing each of the crew to escape. Then, thinking that they had all jumped, he made his way to the emergency exit in the nose and jumped.

De Breyne wasn't to know the rear-gunner Pat Brophy was still aboard. As the hydraulic system was powered by the now useless port outer engine, the hapless gunner could not operate the 'egg' around to get out. He tried to handcrank the turret into a 90-degree position, to enable him to roll out backwards, but the crank broke. Brophy had heard that rear gunners rarely escaped flaming aircraft, and now he knew why. Fighting desperately for his life, he managed to pry the sliding doors open and reach in to grab his parachute pack. The rear gunner accepted that, at least with the bombs still on board, his death would be instantaneous when the aircraft impacted on the French countryside.

Then he saw Andy Mynarski. As the upper turret gunner, Mynarski had had sufficient time to extricate himself from the burning aircraft and jump with the others. But just as he was about to do so, Brophy saw him look towards the rear. Through the smoke and flames, Mynarski must have caught Brophy desperately trying to escape out of his little hell. Mynarski could have saved himself and jumped through the open door right then. Instead, he fought his way back through the corridor, and ignoring the flames that engulfed his own flyingsuit, he grabbed an axe and began to smash at the turret doors. Made to withstand cannon fire, the doors did not cave in easily. By now the Lancaster was in its death dive and as Mynarski wielded the axe, the draft made the plane's interior an inferno. His parachute pack and flying suit in flame and the turret doors as firm as ever, Mynarski realized that he had done all he could. Even then he didn't jump. He stood at attention and in a strange gesture, saluted his comrade. That they were to die in seconds, both knew. Then Mynarski, now a human torch, threw himself out of the aircraft and Brophy awaited death.

To the rear-gunner's amazement, the heavily loaded Lancaster hit the field on an even keel, with the rear gun-turret breaking free on impact. The doors now opened easily and Brophy tumbled out. He got to his feet, soon to be met by French Resistance who had witnessed the whole drama — and the remainder of the crew! Except for Andy Mynarski. The heroic gunner had fallen into a swamp and although quickly rescued by the Resistance, died of his burns. Brophy, still shaken by the death of his crewmate, told the others of Mynarski's self-sacrifice.

Although the crew was soon to be captured and imprisoned by the Germans, upon liberation 6 Group was given a full report of Andy Mynarski's death. On October 11, 1946, he was awarded a posthumous Victoria Cross.

After the war, a Winnipeg high school was named in his honour as were a group of Canadian lakes. The former trapper from Manitoba would have liked that.

Most fitting, the RCAF created the Mynarski Trophy. A handsome silver bowl, it is awarded annually to the RCAF Station whose married quarters community council makes the most effective use of its resources in developing a youth recreation program.

---

1 Thirty sorties had constituted a tour. But now with the bombing of 'softer' targets in France, sorties were given points as to their defences. Sorties over Germany accounted for more points than those over lightly defended targets. A tour was now measured in acquiring a total of 120 points.

# NOORDUYN, ROBERT B. CORNELIUS (BOB)

B ob Noorduyn was born in Nijmegen, the Netherlands, in 1893 of an English mother and Dutch father. Because of his mother's influence, although fluent in Dutch and German, Noorduyn always regarded English as his native language. After a formal education in Holland and Germany, in 1913 he emigrated to England and obtained work at the new Sopwith Aviation company.

From there he moved to Armstrong-Whitworth and much later to British Aerial Transport. While with the latter, he first began designing aircraft.

In 1920 he joined the Dutch aircraft manufacturer Fokker Aviation and came to the United States as assistant to the famous Anthony Fokker. Remembered by the victorious Allies as the designer of deadly German fighters that almost swept the skies in the First World War, Anthony Fokker now concentrated solely on making passenger aircraft. Because Fokker rarely crossed the Atlantic, Noorduyn had a free hand in running the factory and redesigning its aircraft to meet American needs. This would become a source of friction as Fokker had serious reservations about Noorduyn's independence. In time Noorduyn himself began to chafe under the restrictions Holland wanted to impose. Matters came to a head when his conversion of the single-engined Fokker F.7 to the more successful Trimotor met with headquarter's and, especially, Fokker's disapproval.

With youthful zeal and disregard for company hierarchy, Noorduyn pressed on to develop the Fokker Universal. This was an original design of a high-winged, single-engined monoplane that would become a bush plane standard for the next 20 years. Although built primarily for the airline market, because of its strength, the Universal became popular with explorers, miners, and bush pilots. Like the other European-import Junkers, Fokkers were legendary in the 1920s. The DC-3s of their day, they were used in the

Bob Noorduyn's classic bush plane, the Norseman, in a quintessentially Canadian setting. (NAM 3480)

MacAlpine expedition by Dominion Explorers. Northern Aerial Minerals Exploration relied on Fokkers through the Arctic. Garnering even more publicity than either of the first two, a Hollywood movie studio had a chorus line of starlets dance on the wings of a giant Fokker airliner. The Australian explorer Kingsford-Smith used the Fokker Trimotor *Southern Cross* on record-breaking flights across the Pacific. Hardly any Atlantic crossing, and rarely a mineral exploration foray, was flown in the decade without the use of a Fokker.

Noorduyn had gained valuable experience in building aircraft that could operate in extreme conditions with little maintenance. Yet he could not help but notice the number of Canadian bush pilots that came down to the New Jersey factory to buy Fokkers, and listened to them singing their praises. He also felt by now that he had outgrown his compatriot's company.

In 1929, Noorduyn moved to Bellanca Aircraft and was closely associated with the design of their Bellanca Skyrocket. Here too was another versatile aircraft suited for operating in the wilderness. In 1932, he began working for Pitcairn Aircraft. For the two years he spent there, his ideas for building his own aircraft were obviously germinated.

Finally in 1934, he seized the opportunity and moved to the Montreal suburb of Cartierville. Using the old Curtiss-Reid plant, he opened Noorduyn

Aircraft Limited. After years of designing other people's dreams, Bob started with the basic tenet that whatever he built, it would be able to fly equally well with floats, skis, or wheels. His creation was not going to be like a conventional Fokker or Fairchild — wheeled airliners that would take floats and skis as an afterthought. It was to be designed from the first as a bush plane. The float performance was the most difficult to meet, but if the aircraft could be manufactured with this in mind, the wheel or ski performance should be easier. With his new-found Canadian enthusiasm and Dutch background, Noorduyn called the aircraft a Norseman.

Work began on the prototype Norseman in early 1935 and within a year aircraft No.1 — a Norseman Mk.1 CF-AYO, took to the air. The engine used was a Wright R-975-E3 rated at 450 hp for take-off. Although only the developmental version, it required so little adjustment that the aircraft was sold to Dominion Skyways in 1936.

The Mk.II that followed was also powered by the Wright engine, but it became evident that more power was necessary for the Mk. IIIs, so Pratt &

Robert Noorduyn (at the far left) is shown meeting Queen Juliana of the Netherlands in Ottawa during the Second World War. (NAM 14705)

Witney Wasps that delivered 600 hp on take-off were substituted. The Wasp was heavier by about 215 pounds and that would cut into the payload, but the added power made them by far the more popular engine. The buyer was always given the option of either engine.

The Mk. IIIs were followed by the Mk. IV, which in still air at sea level could get off the water in 22 seconds. The bush pilots appreciated the STOL capabilities as it enabled them to get into and out of smaller lakes and rivers.

From the very beginning, the company attracted a dedicated workforce of mechanics, riggers, and fitters at the old Curtiss-Reid plant. It began with a total force of only 30 people but they were soon able to turn out a Norseman every 60 days.

In 1938, the RCMP had one built and in the same year the RCAF placed contracts for eight machines. In view of the gathering storm across the Atlantic, these Norsemen were to be used as bomber trainers and thus were the first to be fitted with bomb racks and sights. By the time the Second World War began, the number of employees at Cartierville had risen to 130, and the plant was being expanded to meet the new government contracts. In 1940 the RCAF placed a huge order for 47 Norsemen to be used as navigational trainers in the British Commonwealth Air Training Plan. The following year, the United States Army Air Force reversed the conventional trend of American aircraft being imported to Canada and placed an order for five Norsemen for its Alaskan bases.

So well did the Americans think of the rugged little aircraft that it saw service throughout their far-flung battlefronts in Asia, Europe, and the South Pacific. In the European theatre, after D-Day, the Norsemen became ambulance aircraft, being able to evacuate the wounded to base hospitals, picking patients up directly behind the front lines. The Mk.VI was known in the USAAF as the UC-64A, and over 700 were built.

Feeling that Victory was near in 1944, with the next series, Noorduyn set aside the "V" for Victory. The "V" Norsemen went into production too late to see wartime service and became known as the Norseman Five instead. In 1945 the Ontario Provincial Air Service purchased the first of the Vs.

The Vs were similar to their predecessors but, with the elimination of military equipment, were 500 pounds lighter. Post-war sales of Norsemen were not brisk. With hundreds of ex-USAAF Norsemen on the market, the aircraft had become a victim of its own success. Of more than 800 Norsemen built during the war, most were flying and available to commercial operators at greatly reduced costs.

The bush pilot's aircraft had become a design classic. Essentially unchanged in aerodynamic detail since 1935, by the 1960s the chunky little aircraft were still flying in over 28 countries. If imitation is the greatest form of flattery, then the Beaver and Otter built by de Havilland in post-war Canada owe a lot to the Norseman.

Individual Norsemen had remarkably long and fruitful careers. The Mk.I prototype CF-AYO saw continous service from its first flight in 1935 until August 28, 1953, when in operation with Orillia Air Services, it crashed in the Georgian Bay area.[1]

In 1946, the rights to build and sell the Norseman were sold to Canadian Car & Foundry, which developed the Mk.VII prototype. During the Korean War, Canadian Car closed down its Montreal operations and shifted the production of Norsemen to its main plant in Fort William, Ontario. In 1953, the company decided not to renew its licence to build Norsemen and disposed of it. The sole Mk.VII prototype had been locked away in a hanger and was eventually destroyed by fire.

Characteristic of his nation, the ubiquitous Dutchman then took back the licence to build his old aircraft and re-formed Noorduyn Aircraft in Montreal to service Norseman users. The last Norseman to be built and test-flown was the 918th, on December 17, 1959. Production was finally halted. Manufacturing an aircraft designed in the 1930s had become increasingly uneconomical. Fabric sewing, doping, and wooden-wing construction were dying arts and the average worker at the Noorduyn workforce was approaching retirement.

Bob Noorduyn didn't live to see the last Norseman take to the air. He died in Burlington, Vermont, on February 22, 1959. But he lives on wherever his stubby little aircraft fly. In 1977 there were still 39 Norsemen with current Certificates of Airworthiness on the Canadian Civil Register. When one considers that their lineage can be traced to the Fokker Universals of the 1920s, this is a wonderful testament to their qualities.

---

1 This same aircraft was used by Warner Brothers in the making of the Second World War film *Captains of the Clouds*. After the movie it flew for such companies as Canadian Pacific Airlines, Cap Airways, Gold Belt Air Service, and Mont Laurier Aviation.

# OAKS, HAROLD ANTHONY (DOC)

Today minerals can be located from the air by the use of airborne scintillometers and electro-magnometers. In the 1920s, 'Doc' Oaks combined the sciences of flying and mining to explore mineral-rich northern Ontario, carrying men and equipment to inaccessible areas. More than any other pioneer, he realized that access to the immense wealth buried beneath the soil there could only be by aircraft.

Harold Anthony 'Doc' Oaks was born at Hespeler, Ontario, on November 12, 1896. To friends, he was known as 'Doc' because his father was a doctor. As much as he tried to get away from this nickname, it remained with him all his life.

Harold joined the Canadian Army in 1915 and transferred to the Royal Flying Corps in 1917. He was soon flying Bristol fighters with No. 2 and later 48 Squadron in France, winning a DFC for gallantry under fire.

After the War he returned to Canada and went to the University of Toronto where he graduated as a mining engineer in 1922. When a university professor gave him permission to do a thesis on the theory of mineral prospecting from the air, Oaks never forgot the experience. He first worked for the Canadian Geological Survey and later the Hollinger Gold Mine, prospecting for minerals in Northern Ontario and Quebec.

In 1924 he was hired by the Ontario Provincial Air Service to fly forestry patrols out of Red Lake. Using a fleet of 13 Vickers HS2L flying boats, the Ontario Provincial Air Service had organized a network of forest-fire patrols over its domain. From the very first day of its inception, the pilots and aircraft flew long, wearisome flights over the dull, green carpet of forest. In 1924 alone over 2,500 flying hours were logged, covering 170,000 air miles, and 597 forest fires were reported. Several of the bush pilots would build up their hours with the OPAS. Among them were Romeo Vachon, Duke Schiller, G.A. Thompson and 'Doc' Oaks.

H. A. 'Doc' Oaks was awarded the Mckee Trophy in 1928 for his work in mineral exploration with the use of an aircraft. (NAM 25927)

Here, Oaks was able to indulge in his two favourite occupations: flying and prospecting. Strikes were being made all the time. In 1925 when the Red Lake Gold Rush began, miners and prospectors flooded the area to stake claims. 'Doc' Oaks and 'Tommy' Thompson could not resist the lure, and that January quit their jobs in the OPAS and travelled by dogsled to Red Lake.

They soon realised that the real money was in transportation and sold their claims to begin an air company. Patricia Airways and Exploration began operations in February with a single Curtiss Lark. This was an open-cockpit biplane, able to carry two passengers. With Oaks as chief pilot, the Lark flew on a schedule between Sioux Lookout and Red Lake. It used floats in the summer and skis in the winter, and by the end of the year it had carried 260 passengers, 140,000 pounds of freight, and 3,000 pounds of mail. As the mail delivery was not government subsidized, a charge of 25 cents was levied for each letter.

Oaks went to Toronto to raise money for a larger aircraft and met the wealthy industrialist James A. Richardson. It was to be a fateful meeting for both.

Richardson had many business interests, but in the early years of aviation in Canada, provided much of the financial backing for bush pilot ventures. Oaks enthusiastically laid down his plans for an airline to fly mail, passengers, and freight in northern Ontario. Richardson agreed to put up the money, and Western Canada Airways came into being in November 1926. That Christmas, Oaks travelled to the Fokker plant in New Jersey and piloted back a Fokker Universal monoplane. He brought it through a blinding snowstorm. Although it could carry five passengers in enclosed comfort, the pilot was still in an open cockpit.

The acquisition of the Fokker could not have come at a better time. The Department of Railways and Canals were planning to build a railway terminal at the port of Churchill on Hudson Bay. The port was ice-bound for half the year and surrounded by miles of frozen or near-frozen wasteland. The railway ended 280 miles away, at Cache Lake, and all equipment, spare parts, and personnel would have to be flown between it and the port.

The department approached Oaks as general manager of Western Canada Airways and asked if air freighting on such a large scale was possible. In February 1927, Oaks accepted the contract and went back to Fokker's to buy two more Universals. Western Canada Airways then flew 27 round trips between Cache Lake and Churchill carrying over eight tons of equipment and fourteen engineers. The operation was a huge success and opened the door for future aerial development of the North.

Servicing aircraft in extreme winter conditions required the construction of characteristic "nose hangers" — these were small tents or frame structures about 12 feet square. The aircraft was drawn into the enclosed area and the curtains were fastened around the bow. A stove was put in the middle to keep the oil and mechanic warm. The whole hangar was on skids, for easy movement. Created by Oaks, these nose hangers remain commonplace in the North even today.

As a former prospector, Oaks realized that Western Canada Airways could service the mining camps now springing up in Manitoba, and Saskatchewan as well as northern Ontario, and he extended the network accordingly.

An excerpt from a 1928 issue of the *Northern Miner* by Norman C. Pearce, illustrates the type of flying done: "In the winter they take a couple of prospectors away north to the centre of their operations. They carry in their season's supplies and equipment, including a canoe that breaks down into three pieces ... and an outboard motor with a supply of gasoline. The prospectors camp until the snow goes, and then have the valuable breakup time for their work which gives them a six weeks jump on their competitors."¹

By 1929 Western Canada Airways had received a post office contract to operate a weekly airmail service to serve the Cold Lake mining area, an established concern.

Oaks was the first to be awarded with the Trans-Canada (McKee) Trophy in 1928. The award was made by the Minister of National Defence, the Honourable J.L. Ralston, on the basis that no one had done more to organize an aerial network in the North, to supply outlying mining communities.

Oaks wanted to concentrate on more intense aerial exploration and once more succeeded in convincing Richardson to back him in a new venture. Northern Aerial Minerals Exploration was exclusively an aerial prospecting company, operating from Sioux Lookout. Extensive planning and investment was involved. Flying in such remote country meant that fuel caches had to be sent to designated points months before, and canoes and food made available for the geologists that the pilots would bring out.

In August 1928 Oaks made his way once more to the Fokker factory to buy another Universal. This time he flew from New York to the Yukon. Like most prospectors, he had heard of the fabulous lost McLeod Mine (and the murder story that went with it) somewhere on the Nahanni River banks, and he set out to look for it.

Through what was left of the summer, Oaks and a prospecting party hunted for the mine, but by September when the snowstorms began, he had to return to civilization without any success.

That December, it looked as if the bush-pilot's luck had run out. He was flying a pair of vacationing honeymooners and two missionaries to Rupert House on the eastern side of Hudson Bay. Without warning a blinding snowstorm hit the aircraft and with visibility at zero, Oaks missed Rupert House landing strip altogether. He landed on rough sea ice and broke a ski. That night, they heated their rations with a blow torch and sheltered in the aircraft cabin as best they could. Lost, they realized that no one knew where they were. The next morning with the temperature at -40 F and the blizzard still raging, Oaks set out to find Rupert House on foot. Fortunately, he recognized a landmark and on returning to the aircraft met an Indian trapper who relayed the information to the settlement. That night, New Year's Eve 1928, a dogteam arrived and rescued the party. The following day, a blacksmith repaired the ski and Oaks took off for home.

More than anyone else, it was the lone prospector that appreciated the bush pilot the most. All summer long, he would pan bars and dig away, beating off mosquitos and sleep under canvas. In the days before radio, the sense of isolation must have been especially hard to bear. Bush pilots recalled that often, in desperation for some company, a miner or trapper would cut spruce boughs or form rocks into distress signals, hoping that an aircraft would fly over. If a pilot did land, he would be greeted by a prospector with the beard of a Biblical prophet and a wild look in his eye. Questions would pour out: "What day was it? What month was it? What had happened in the world? Did he have any sugar? Or tobacco?"

Through those early years, Northern Aerial Minerals Exploration (NAME) acquired more aircraft and established 33 fuel caches in the North. They were thus able to put a paying passenger anywhere within a short time. But for what the prospectors could pay, coupled with the price of transporting fuel to distant caches, it was not financially viable, and NAME was forced into bankruptcy. Then, too, the hazards of long-distance flying were still very evident both to the public and the company. In August 1929, one of the NAME aircraft ran out of fuel and was forced down on the Arctic mainland coast. Although the crew and, eventually, the aircraft returned, it was a sobering reminder.

Other dangers were the unreliability of maps in an area that was almost unexplored. Navigation was completely by sight and often the location of rivers needed for direction had no relation to their positions on maps. Only the Inuit had a practiced eye for recognizing geological landmarks, and for this reason pilots often took them onboard as navigators. Owing to the nearness of the Magnetic Pole, compass readings were unreliable. Landing on lakes, bays, or even rivers was especially dangerous for a floatplane, as the depths and location of rocks were unknown.

By 1932 the prospecting operations were becoming increasingly cost-ineffective and NAME sold most of its equipment and routes to Canadian Airways. In 1936 it ceased to exist altogether. 'Doc' had left the company by then to establish Oaks Airways in 1931, still at Sioux Lookout. His company operated a Fairchild KR-34 and a Junkers W-34, but it was not a commercial success and other bush pilots took over the routes he had pioneered.

During World War II, Oaks worked for his friend Clark Ruse in Nova Scotia, returning to mining afterwards at Port Arthur. In 1945 he published his treatise, "Function of the Aeroplane in Canadian Prospecting," in a mining journal. In many ways it was an explanation of his life's work. In 1953, he became a consultant for the James A. Richardson Company. He died in Toronto on July 21, 1968.

---

1 Sutherland, Alice Gibson, *Canada's Aviation Pioneers* (Toronto: McGraw-Hill Ryerson, 1978), p. 24.

Pat Reid beside his Fairchild during the search for the American explorer, Carl Ben Eielson, November 1929. (NAM 4748)

# REID, THOMAS MAYNE (PAT)

Hollywood could not have scripted it better. A bush pilot searching for a downed aircraft, in a raging snowstorm, is himself forced down into a narrow mountain pass. As the pilot tries to land the plane on a creek bed, the mountain walls rip off the aircraft's wing-tip, and he is marooned. A week later, not only does our "hero" repair the wing but threads his way out of the mountain valley and subsequently flies over the coast of Siberia to search for the lost plane.

Thomas Mayne Reid was born at Ballyroney, County Down, Northern Ireland, on August 22, 1895, and trained as an engineer at the Ferguson Automotive Company in Belfast. In 1915 he joined the Royal Naval Air Service and served in the Dardanelles campaign as an engineer in a blimp. In 1917, he was posted to Dunkirk, France, to be trained as a pilot in a flying boat. He joined the North Sea patrol in the final year of the war on anti-submarine duties and was awarded the Distinguished Flying Medal for gallantry.

When the war ended, he worked for the Handley Page Transport Company at Croydon, England, then flying between London and Paris and, later, Brussels and Amsterdam. He emigrated to Canada in 1924 and immediately secured a job in the Ontario Air Provincial Service as a pilot. He joined H.A. 'Doc' Oaks and Jack E. Hammell in 1928 with their Northern Aerial Mines Exploration (NAME) concern, based in Sioux Lookout, Ontario. This was a mineral exploration company that transported all the prospector's food, fuel and equipment by air.

In 1929, with two other pilots, Reid flew a party of prospectors from Fort Churchill, Ontario, to Coppermine on the Arctic Ocean, then down the Mackenzie River to Edmonton. He piloted an ungainly looking Loening amphibian that was destined to run out of fuel on another flight in the tundra later that month. The whole expedition lasted six months and Reid crossed 25,000 miles of barren wilderness without the benefit of navigational aids. The pioneering of the Northwest Passage by air made this an historic flight.

But Reid is ultimately remembered for his daring rescue attempt of the

American explorer, Carl Ben Eielson. In November 1929, Eielson, already famous for flying the first Alaskan airmail, set out to rescue the crew of a schooner that had been stranded for three months off the Siberian coast. On his first flight, Eielson was able to take half the crew off the ship but on the way back, he ran into stormy weather and disappeared.

When Eielson had not returned, search parties were sent out to find him. American and Soviet fliers scoured the inhospitable region. Pat Reid, a veteran of the MacAlpine Expedition rescue, was hired by the Aviation Corporation of New York to lead a rescue attempt. Flying NAME's Fairchild, he and his crew took off from Fairbanks, Alaska, only to be caught in a heavy blizzard, which prevented them from flying over the mountains. Reid was now hemmed in by the ever-narrowing mountainsides as he looked for a pass through them. Finally, he chose to land the aircraft in a creek bed and wait out the storm.

But as he concentrated on getting the Fairchild down, the valley walls were so tapered that the right wing-tip hit a ledge and was broken off. The rescuers themselves were now stranded. Faced with the choice of either repairing the wing and trying to fly out or putting on their snowshoes and risking death in the mountains, Reid and his crew chose the former. By sleeping in the plane for a week and sharing their emergency rations, they managed to repair the wing-tip and scrape through the valley into the air. Reid arrived back in Fairbanks and had the strange experience of seeing his own obituary in the local newspaper.

They continued their search for Eielson's aircraft, flying over the Bering Strait and along the Siberian coast. On January 30, 1930, two of Eielson's friends in a pair of open-cockpit aircraft, sighted what they thought was the sail of a schooner on a frozen lagoon at the mouth of the Anguema River in Soviet Siberia. It was the wing of Eielson's aircraft and they, with Reid, landed to search for the remains of its pilots. When the bodies of Eielson and his mechanic were found under eight feet of snow, Reid flew them to Fairbanks and later Seattle, Washington, for burial.

An inquiry into the crash concluded that the aviators had plunged into the ocean because of the phenomenon most dreaded by all bush pilots — the total whiteout. Still feared today, a white-out is snow below and white drifting fog all around. Disoriented and disbelieving his instruments, the pilot cannot tell whether he is in a climb, a slip, or a stall.

To Pat Reid, the North and aviation were completely interlocked. He made pioneer flights to the Arctic coast, Great Bear Lake, the Mackenzie River area, and the eastern coast of Hudson Bay. He spent several nights waiting out storms in the shelter of his aircraft cabin.

The bush pilot joined Imperial Oil in 1931 as their aviation manager and flew his company's entry, a de Havilland Puss Moth, in the Trans-Canada Air Pageant. This was a transcontinental tour that started in Hamilton on July 1 and toured every major city in Canada such as Halifax, Quebec City, Montreal, Edmonton, Vancouver to end at the Canadian National Exhibition in Toronto on September 8. Reid also took the Moth down to the United States on a goodwill tour on behalf of the de Havilland factory in Toronto.

Imperial Oil sought to become the major aviation fuel supplier in the Dominion, and as early 1921 had operated a fleet of aircraft of its own. By hiring Reid, with his wide experience in bush flying, the company was able to expand its empire from major airports like Malton to a network of tiny bush communities.

Always one to spot a publicity opportunity for his employers, Reid was on hand in 1933 at Shediac, New Brunswick, to welcome the Italian Air Armada of 24 twin-hulled flying boats, led by the flamboyant Italian General Balbo, which were enroute to the Chicago World's Fair.

By 1939 Reid's distinctive Beechcraft biplane had become a familiar sight in communities from the Labrador coast to Aklavik. As a good friend of the aviation historian Frank Ellis, he was able to keep the latter up-to-date on all historic developments.

Twice during the Second World War, Reid was awarded the McKee Trophy for his part in organizing, with Imperial Oil, the national fuel supply for the military. He was still employed by the fuel company as manager of their Aviation Sales Department when, in 1954, he met his death tragically. On April 8, the 25th anniversary of his historic Northwest Passage flight, the Trans-Canada Air Lines aircraft that was carrying Pat Reid and his wife collided with an RCAF trainer over Moose Jaw, Saskatchewan. Pat Reid was not one of the survivors.

T.W. Siers, maintenance manager for Western Canada Airways, awarded the Mckee Trophy in 1940 for the development of the oil dilution system.
(NAM 25045)

# SIERS, THOMAS W. (TOMMY)

A pioneer aviation engineer, Tommy Siers is gratefully remembered by bush pilots for his perfection of the Worth oil dilution system. The problems of starting a frozen aircraft engine in the Arctic, when flying time is limited to a few daylight hours, had plagued the early pilots and retarded the advance of commercial aviation north of Edmonton. Tommy Siers's solution was a major contribution to the development of aviation services in cold-weather flying.

Siers was born on May 13, 1896, in Yorkshire, England. After completing his education there, he emigrated to Canada, beginning as an apprentice machinist in the Canadian National Railway shops in Winnipeg, Manitoba.

The aircraft used in the search for the MacAlpine party, Cranberry Portage, 1929. (NAM 2311)

In 1914, he took automobile engineering courses at the Technical School of Winnipeg. Siers joined Lord Strathcona's Horse in 1917 as a trooper and was posted to Shornecliffe, England. He saw action in the final year of the war at the Battle of Arras, returning to Winnipeg in 1919. A year later, he signed up with the newly formed Canadian Air Force at Camp Borden and learned to overhaul aircraft engines.

In 1920 Siers left the air force and went a familiar route for those seeking a career in aviation in those days. He joined Laurentide Air Service at Lac a la Tortue, Quebec, as an aero engine fitter. As there were no airlines then, the forestry patrol companies served as schools for many aspiring bush pilots after the First World War. Later, Tommy Siers overhauled engines for the Ontario Provincial Air Service.

In 1927 he began working in Winnipeg for Western Canada Airways as the chief mechanic. His first taste of winter flying occurred in 1929 when the company secured the contract to transport the MacApline expedition to the far North. Although Sier's part came much later, the drama that unfolded clearly illustrates the hazards of flying in the North.

With a small crew, Siers was the engineer in charge of keeping the expedition's Fokker and Fairchild flying. In September 1929, Dominion Explorers sponsored Lieutenant Colonial C.D.H. MacApline to explore the coast from Churchill, Manitoba, to Aklavik, along the Arctic Ocean, and return with them.

The expedition was dogged by misfortune from the beginning. Their supply ship, the schooner *Morso* on the way to Churchill, caught fire in Hudson Bay. The conflagration ignited the dynamite in the hold and the resulting explosion sent the ship and all the expedition's fuel and food to the bottom. When the aircraft — now the only means of contact with the outside world, landed at Churchill harbour, the streak of bad luck continued. The Fokker, moored offshore, dragged her anchor one night and was carried out to sea by the strong tide. Inevitably caught by sharp ice floes, she sank within sight of the rescue party. A replacement Fokker was sent from the Western Canada Airways base in Winnipeg, but by now the expedition was a fortnight behind schedule and winter was closing fast.

Attempting to make the best of the situation, MacApline ordered that both aircraft begin flying as far as they could with the little fuel they had left. Violent storms pushed the pilots further north than they had planned and, almost out of gasoline, they reached Dease Point, on Queen Maude Gulf. Using their last drops of fuel, the pilots flew in ever-decreasing circuits looking for the Hudson Bay post that friendly Innuit said was nearby.

It was in vain. Without radio communications or gasoline, the expedition was now stranded. At first they planned to wait out the winter for rescue, but when supplies soon dwindled and the intense Arctic conditions made themselves felt, they decided to walk towards civilization. As soon as the ice was thick enough, using dogteams and food given by the Innuit, the expedition set out to walk to the nearest village. Almost two months later, at 7:30 a.m. on November 3, 1929, MacApline and his men saw the lights of the Hudson's Bay store at Cambridge Bay. They were suffering from starvation, frostbite and exhaustion. All of them realized that had it not been for the generosity of the natives, they would have long since met the same fate as the Franklin expedition of 1831, almost a century before.

The rescue effort that Dominion Explorers mounted almost met with disaster itself. When landing at Bathurst Inlet, one Fokker plunged through the ice; only its wings prevented it from sinking completely. Aware of the coverage that two aircraft could fly in the search, instead of only one, Tommy Siers and his crew had the machine raised from the ice by block and tackle and, in spite of the freezing conditions, overhauled the engine in record time. Within 10 days, although a blizzard raged about them, the Fokker was ready to fly. Both aircraft soared over Dease Point, but by now the marooned party had left to make its way to Cambridge Bay.

The second aircraft was then forced down at Musk Ox Lake because of low fog. Its skis became damaged by the huge snow drifts as the pilot attempted to take off when the weather cleared. Again, trying to keep both planes in the aerial search, Tommy Siers in a superhuman effort repaired the undercarriage, escaping land before the engine froze completely. When they returned to base, word was received that the expedition had made it to Cambridge Bay safely.

From this series of misadventures in the Arctic, Siers gained a healthy respect for the harshness of the northern climate and its effect on aircraft engines. In 1931, he visted the Junkers aircraft plant at Dessau, Germany. Here, Hugo Junkers was building his 'Tante Ju' that would become familiar to a generation of Canadian bush pilots - the indestructible Junkers 52. Siers took the opportunity to study the German engines and later made a tour of the de Havilland and Noorduyn factories in Canada as well.

Because of the limited market, commercial aircraft were at that time not specifically designed for the Canadian market and imports had to be modified to operate in the cold climate. Sier's experience with the MacAlpine expedition, coupled with his mechanical flair, made him think of adapting the aircraft models available to better suit Canada's rugged conditions. His work on modifying skis, ski pedestals, cabin heaters, and carburetors all hastened the spread of reliable, cheap air services in the Arctic.

He is, however, best remembered for the exploitation of the Worth oil dilution system. From the earliest days of bush flying, pilots had to heat the oil in aircraft engines at dawn in order to start them in subzero temperatures. This was usually accomplished with a blowtorch, or with wood stoves placed under the engine, or if the power was available, an electric heater. Whatever method was used it was slow — depending on the frigidity of the air and wind - and dangerous. As soon as a pilot landed his aircraft for the night, for no flying was done after sunset, a ritual was observed. His engineer immediately drained the oil out of the hot engine. Because of the cold, the oil would freeze solid as soon as it fell on the ground, and the crew took the fan of black frozen spray into the tent that night. The next morning, hours before it became light (for with the few hours of daylight to fly in every minute was precious) the engineer heated the oil on a stove, laboriously stirring it until it became liquid. Then he poured it back into the engine and prayed that it would start.

Commercial operators realized that they could never operate profitably by flying in these conditions and sought some way to prevent the oil from freezing overnight. On December 15, 1935, when a Junkers belonging to Canadian Airways fell through the ice on the lake at Fort Chipeywan, Alberta, Siers was called to the scene. In -60 degree weather, he raised the aircraft out of the ice and set to repairing its Pratt & Whitney Wasp engine. Because of the gasoline escaping from the engine, a woodstove or blowtorch could not be used, and the resulting discomfort and frostbite suffered by Siers can only be imagined. However, this was an opportunity to test his theories on starting aircraft engines at subzero temperatures. What he passed off as his 'tinkering' worked, and soon the Junkers was flown off the lake.

In 1937, Tommy Siers, as superintendent of maintenance for Canadian Airways, began to investigate adapting the United States Army Air Corps oil dilution system to Canadian aircraft. This had been invented by Weldon Worth for the USAAC and tested on military aircraft at Dayton, Ohio. The system was based on the principle that engine oil when diluted with the addition of gasoline would retain its lubrication. The thinning would allow the engine to start up immediately, even after exposure to below-zero temperatures.

Siers was amazed to learn that although the Worth system was known in aviation circles, nothing had been done to adapt it for commercial use. In 1938, he designed the system's 'tank-within-a-tank' and installed it on one of Canadian Airway's Noorduyns. In what seems to be a pattern in his life, bad luck in the form of thin ice, struck Siers again. The Noorduyn, equipped with his oil dilution tank, started up in freezing conditions and took off on the ice at Bill Joe Lake, Ontario. Unfortunately before Siers had an opportunity to examine the engine, the aircraft made a rough landing and crashed through the ice to sink into the lake. It was to be early 1939 before the oil dilution tests could continue, but by now Siers had ample evidence that the system worked. Even with the temperature at -35 degrees F., the engines he used fired up immediately.

In 1940 he put his modifications down on paper and published them. The inventor, Weldon Worth, agreed that it was due to the Canadian's pragmatism that his invention was properly exploited. In the winter of 1940-41, every Canadian Airways aircraft engine was modified to include the Sier's oil dilution system. Even in temperatures as low as -57 degrees F, all planes started without hesitation.

This was especially fortunate given the onrush of world events at that time. The Allies were now operating as far afield as Greenland, the Soviet Union, and, closer to home — Alaska. Strategic air operations were being flown in Greenland's inhospitable climate, and lend-lease bombers were being ferried to the Russians across the icy Bering Strait. Siers provided full details of his oil dilution system to the American and British governments, and, in Canada, to the National Research Council and Department of Transport.

In 1941, Tommy Siers was awarded the Trans-Canada Trophy for his development of the Worth principle of the oil dilution system for cold-weather starting of aircraft engines. Through the war and after, he worked for Canadian Pacific Airlines, only retiring in 1961.

Of all the tributes he received for his creation, none conveys the drudgery of starting a cold engine better than this anonymous poem.

**An Engineer's Winter Song — Sioux Lookout**

Six a.m., the clock never freezes,
Everything else does by Jesus
Dark as pitch and cold as hell
Why I'm an engineer I can never tell.

Light the fire and a cigarette,
Get the torches and oil pail set,
Out in the cold at forty below,
Under the cover for an hour or so.

Cold and alone while the fire pots roar,
While in the shack the pilots snore
Snug and warm in their sleeping bags,
Trying to recover from last night's jag.

In with the slippings, put the cover on the back,
Wind the inertia 'til the snow turns black,
Jump in the cockpit before she runs down
And wind the booster around and around.

I thought it would be romance and fun
Looking after a Fokker seventy-one,
Freeze your face and your fingers too
Servicing a Wasp north of fifty-two.[1]

1 I am indebted to Stan Knight for this poem by an anonymous engineer, published in the
CAHS *Journal,* Fall 1972.

# STINSON, KATHERINE

Even today, stunt pilot Katherine Stinson would be considered remark-able. In an era when members of her sex were not seen as mechanically minded enough to drive a motor car, let alone fly an aircraft, she demonstrated otherwise. In two summers, she proved that the new science of aviation was not exclusively a male preserve.

As late as 1915, few outside Toronto, Montreal or Vancouver had seen an aircraft. While the newspapers were full of the air duels over the Western Front, to small communities in most of Canada, pilots and their machines were still part of a circus sideshow, providing thrills and excitement.

For aviators trying to earn a living when commercial aviation did not exist in Canada, this meant flying wherever outdoor crowds met — fairgrounds, parks, and agricultural exhibitions. Dollar-a minute joy rides, stunt flying, and wing walking were the standard fare of barnstormers.

The most unusual of this breed was a quiet, unassuming American teenager called Katherine Stinson. Born in San Antonio in 1899, she first flew at the age of 12. She earned her Aero Club of America pilot's licence in 1912 and with her brother and sister founded an aviation school to train North American pilots for the Royal Flying Corps. She, herself, volunteered her services as a trained pilot to the Allies but was refused because of her sex.

At her Stinson School of Aviation in San Antonio, Texas, she was amazed by the number of air-minded Canadians who travelled so far for flying lessons. By 1916, she had taught a total of 60 Canadians to fly — some of whom were now in action over the Western Front. Again she tried to sign up with the Royal Flying Corps and follow her pupils to France, but was once more refused. When her grateful graduates encouraged her to tour Canada with her aircraft, she agreed.

The only organization that saw the value of an aviatrix for publicity pur-poses was the Red Cross. They sponsored her on record-breaking flights across the United States and Canada in their name. In 1917, she flew from Buffalo to Washington, a distance of 670 miles in 11 hours and 25 minutes to hand

Katherine Stinson beside her Curtiss biplane, July 1918. (NAM 2649)

over sponsored cheques of over $1 million to the president. Even then she was refused permission to fly over the Western Front. In vain she volunteered to drive an ambulance, but the intolerance of the time proved too strong and she turned to barnstorming to earn a living.

Katherine Stinson first appeared on the prairies in 1916. Her celebrity status assured large turnouts at airshows, and no shortage of commercial sponsors. Beginning June 30, she thrilled the crowds that gathered to watch her in Calgary's fairgrounds. She flew three flights daily for four days, turning loops, spirals and dives. The wartime boom in aircraft manufacturing had not made its presence felt in Alberta, and aircraft were still home-made and expensive. Her first aircraft was a second-hand biplane built by an American inventor Matte Laird. Its seven-cylinder rotary Gnome engine had been taken from a fatal crash a year earlier. Both served Miss Stinson well, and that summer she was feted by the public and military alike.

Nothing daunted the young American. Miss Stinson flew in exhibitions at Edmonton, Brandon, Regina, and Winnipeg. At Brandon, her flying so impressed the local Sioux tribe at the rodeo that its chief made her an honourary princess. Yet the very next day her celebrity status did not prevent her from being arrested as a spy by the military for landing unannounced at Camp Hughes. Like a heroine from the silent screen, she flew in unannounced, landing on the parade ground, expecting an invitation to lunch. Instead with the ever-present German spy scare making the rounds, she was met by soldiers with fixed bayonets and escorted to prison.

The newspapers of the day note that she specialized in a heart-stopping dive. So realistic was it that at Regina, when she performed this stunt, the city authorities were inundated with telephone calls to ask where she had crashed.

On August 3 in Winnipeg, it was advertised that she would fly to raise money for the Patriotic Fund. But assembling her aircraft took longer than expected, and she was not ready until the next day. August 4 dawned murky and black, but the crowds poured in to watch. When she made the decision to fly at 10:30, the sky was threatening and the onlookers feared for her safety. But she took off into the dark sky with characteristic courage. The owners of motor cars were asked to park around the fairground boundaries and switch their headlights on to guide her back. Soldiers were posted with flare guns to provide help if needed. Concerned citizens lit bonfires at either end of the field.

Stinson flew into the gloom and at an altitude of about 1,000 feet, she began her show. She ignited flares attached to the rear edges of the wings and illuminated the aircraft's shape to the crowd below. Cheers rang out as she circled over the grounds, dodging and weaving through the overcast. When her flares finally died out, she cut the engine to a low hum and landed in the centre of the field. The audience was now delirious.

On the last day of the exhibition, to the delight of the crowds, she 'bombed' a dummy fort that had been built on the grounds while its defending soldiers fired blank cartridges at her aircraft.

That winter she claimed to have toured China and Japan with her aircraft and crew, creating a sensation in those societies not only as a woman who flew but as one who flew in trousers!

In 1918, Katherine Stinson returned to charm Canadians with her barnstorming once more. Now using a war-surplus Curtiss biplane, she again stunt-flew at the exhibition in Calgary. But her mind was now on more serious feats.

Contracted to perform before fairgoers in Edmonton, on her flight there she agreed to take a small bag of mail. In it there were exactly 259 letters, each of which bore the stamp "Aeroplane Mail Service, July 9th, 1918, Calgary, Alberta" in violet ink. Miss Stinson took the bag from the postmaster and left at 1:30 p.m. for the 175-air-mile flight to Edmonton. Along the way, the engine gave her trouble and she was forced to land to make repairs. At 8:00 p.m. that evening, anxious supporters watched her land at Edmonton, taxi up to the postmaster, and hand him the bag of letters. At age 22, she had set a record, for her actual flying time was 2 hours and 5 minutes.[1] More importantly, this was only the second air mail-flight in Canada, the first being the month before. She had proved to the post office the viability of air mail.

In Edmonton she also showed that she was not only an excellent aviatrix but a racing car driver as well. While at the local fair, she set a speed record in her Fiat for a mile: one minute, fifteen seconds. Typically, she continued on with her barnstorming contract, giving air shows in Saskatoon, Red Deer and Camrose.

Stinson went on to tour as much of Canada as possible that summer. She stayed in Ottawa for a week to visit families of former pupils and she flew at Toronto's Canadian National Exhibition.

The *Peterborough Examiner* breathlessly recorded that the world-famous aviatrix would be "stunting" (weather permitting) on September 12th at the Industrial Exhibition in their town. When interviewed by the *Examiner*, Miss Stinson said that she was planning to tour Cuba and South America as soon as she left Canada. When asked if she was afraid of flying, she replied that she had never been and had only gotten into this business to make money for her musical education.

September 12th and 13th were wet and windy and the large crowds that gathered to see her 'stunt' were disappointed. In spite of an uncomfortable drizzle the next day, the 14th, she had the machine towed out of the hanger and at 7:00 p.m. took off across the sodden field. Those that came saw her swoop down, twist, and turn under an overcast sky.

Just as she had at exhibitions on the prairies, Katherine Stinson entertained the audience for several shows. In an age when flying was associated with burly males, dirty overalls, and castor-oil fumes, Miss Stinson always alighted from her aircraft as the newspapers observed "as pretty as an angel." The exhibition ended the next day and the aviatrix left Canada forever on September 15th.

For many years it was rumoured that Stinson's biplane had been sold to a local flying company, Canadian Air Services, based at Harwood, Ontario. Was she bankrupt at this point in the tour or having 'stunted' for six years, knew that she was running out of luck? We will never know as the Peterborough exhibition was her swan song.

The aviation historian Frank Ellis wrote that, after it, Katherine Stinson married and went to live in Santa Fe, New Mexico. She disappears from the pages of history at this point. She had struck a blow for feminism along with Amelia Earhart, Amy Johnson, Elsie MacGill, and Jackie Cochrane. But to the Canadian public in the early years of this century, she must have seemed like an angel come down to earth for a few moments, heralding the Air Age.

---

1 Ellis, Frank H., *Canada's Flying Heritage* (Toronto: University of Toronto Press, 1954), p. 135.

J. H. 'Tuddy' Tudhope at Camp Borden, 1921. (NAM 10179)

# TUDHOPE, JOHN HENRY (Tuddy)

During the 1920s and 1930s, 'Tuddy' Tudhope pioneered mail routes across the country by carrying out aerial surveys for the building of the Trans-Canada Airway. As superintendent of airways with National Defence, he personally travelled from Cape Breton to Victoria by rail, foot, and sometimes air, setting out air fields that would allow the early trans-Canada mail and passenger services to operate safely.

John Henry Tudhope was born in Johannesburg, South Africa, on April 17, 1891, and served in the German Southwest African campaign until 1915. Then he joined the Royal Flying Corps and found himself with 40 Squadron over the Western Front. He was awarded the Military Cross and Bar for conspicuous gallantry in attacking enemy aircraft. Retiring from the RFC as a major in 1918, he tried farming in British Columbia for a year. But the lure of flying was too strong and he enlisted in the nascent Canadian Air Force to be posted as an instructor to Camp Borden in 1920. He specialized in seaplane flying and became known as 'Tuddy' wherever he flew.

After a short spell of flying forestry patrols for Laurentide Air Services in Quebec, he returned to the RCAF in 1924 to be posted to Jericho Beach Air Station, Vancouver. A year later, he left as squadron leader to become superintendent of airways with the Department of National Defence on July 1, 1927.

It was here that Tuddy hit his true metier. In 1927, the post office was considering ways of speeding up the delivery of trans-Atlantic mail. There might be little that could be done about flying the vast Atlantic but if seaplanes could meet the ships that carried the mail as far out into the St. Lawrence as possible, the mail could then be flown efficiently on to Montreal. Tuddy's orders were first to survey the North Shore of the St. Lawrence and look for sheltered air harbours to fly from. He was then to attempt a mail transfer at sea himself. Flying a Vickers Vanessa floatplane on September 9,

1927, he took off from Rimouski harbour to meet the incoming *Empress of France*. Unknown to him, because of high seas the night before, the landing struts on the Vanessa's pontoons were weakened. Just after dawn, Tuddy met the pilot boat in the middle of the St. Lawrence and took on board the 500 pounds of mail from the *Empress*. Then everything went wrong. As he taxied to take off with the heavy bags in the choppy seas, a strut gave way and the port wing tipped into the water. This caused the propeller to rip into the forward section of the pontoon, and cut it in half. Fortunately, the pilot boat was still within sight and quickly came to the rescue to tow the crippled floatplane back to Rimouski. The aircraft was declared a wreck and the mail arrived in Montreal by train long after the *Empress* did.

Flights were made to other incoming ships through that autumn, with some variation. Eventually, practicality won over drama, and the mail was carried by pilot boat to Rimouski airport where Tudhope flew it in a wheeled aircraft to St. Hubert, near Montreal. By 1928, the Rimouski-Montreal airmail route could be operated by a commercial airline, the flights being made to coincide with incoming and outgoing ships.

Pleased at Tudhope's work, that same year a decision was taken by Ottawa that would affect the future of Canadian aviation, and Tudhope's life in particular. Worried about the incursions of American airlines flying into the Canadian West, the government launched a program to survey and construct the facilities for a Trans-Canada Airway over which reliable mail service could be operated. The whole scheme, comparable to the construction of the Canadian Pacific Railway in the last century, was to be completed by July 1, 1937. The selection of aerodromes from Winnipeg to Halifax was to be done by Major Tudhope, Robert Dodds, and George Wakeman.

The three pilots divided up the huge, almost unexplored expanse of country that they were to survey. Dodds chose the area between Winnipeg and Kapuskasing, Tudhope from Kapuskasing to Montreal, and Wakeman from Montreal to Halifax. All agreed that the most difficult sites for airports would be the bush country between Cochrane and the Manitoba border. But Ottawa estimated it would need at least 27 airports, built on the muskeg and rock, on that section.

Tudhope followed the railway lines from Ottawa to North Bay and then to Cochrane. He noted that there were many lakes which would be excellent for floatplanes, but few suitable landing sites for wheeled aircraft. For detailed inspection, aircraft could not be used, and it was either by foot or the Ontario Forestry Branch's 'gas speeder' — a gas operated railway car — that the airmen used to look at the thousands of miles of terrain.

By 1930 he had surveyed an airway route from Ottawa through the

Praires and Rockies to Vancouver. Over the praires he was able to fly a variety of aircraft — a Pitcairn Mailwing G-CAXJ, a D.H. Moth, and a Stearman. The most spectacular leg was undoubtedly through the Rockies. Tudhope made a preliminary air survey from Lethbridge through the Crowsnest Pass to Vancouver. Much valuable data was obtained, that ranged from the topographical features for landing fields to meteorological conditions. However, with the beginning of the Depression, the ambitious plans for a cross-Canada air route had to be shelved, and parts of the program were not to be implemented until the Second World War.

For efficient airmail connections, night flying would have to take place. This meant 24-hour airports that were equipped with radios and boundary and obstruction lights to guide in aircraft. Tudhope and his team intensified their efforts, not only in the West but also between Windsor-Toronto and Montreal-St.Jean-Moncton.

Building an all-weather, lighted airfield in the middle of the bush required a preliminary aerial reconnaissance at low level to get an idea of the ground conditions. When a highway, a railway line, or a lake did not exist nearby, the survey party hiked from the nearest railhead to make a ground survey. If the site met the requirements, then applications were made to the relevant authorities: the timber companies, the federal or provincial governments, the Department of Mines, or, if it was privately owned, an option to purchase was obtained from the owner.

If this was successful, then a legal survey was made of the site and it became the property of the Department of Defence. Supplies and men were brought in first on foot and then later by horse and tractor. After cutting a fireguard around the site to prevent forest fires, the land was drained if it was on muskeg or burned if the muskeg was too deep. Rocks and boulders were hauled away. Runway layouts were staked and marked. Brush and timber were cut and burnt. Using a caboose supplied by the railway, and sometimes a gas speeder, the inspection teams arrived to make constant surveys of the airfields at various stages of development.

The workers — recruited during the Depression from the growing ranks of those on relief — used teams of horses and dynamite to clear the tree stumps, and wheelbarrows to carry gravel in to smooth the field. The work was hot and backbreaking, and the teams suffered black-fly plagues throughout the summer. The provision of the airport caretaker's facilities required that power lines be laid to the camp or, if this was not possible, generators and fuel depots built. The caretaker's cabin, hangars, and meteorological and radio facilities were gradually added, as were the electric beacons. Obstruction lights, runway and perimeter lights marked the airfield, and radio antennae were installed. The building of the Trans-Canada Airway was akin to the con-struction of the trans-continental railway of another era and the Distant Early Warning line of the future.

On March 26, 1931, in recognition of his outstanding work in carrying out aerial surveys, Squadron Leader Tudhope was presented with the Trans-Canada Trophy by the minister of national defence. The ceremony took place at the Chateau Laurier in Ottawa before a gathering of First World War air aces.

By now, with the Depression biting deeply into the government's budget, aerial surveys were cut back drastically. The only reason, it sometimes seemed to Tudhope, that the trans-Canada airway program was even considered was because it provided employment for the hundreds of men now available to construct the airports.

As inspector of airways for the Department of National Defence, Tuddy continued to travel to the sites — by air, by rail, and on foot. He charted the best routes for the longer-range aircraft that were now coming on line. Flying a Fairchild FC2 Razorback, he surveyed routes from Winnipeg to Toronto, via North Bay. Kapuskasing, South Porcupine, Kowkash, and Nagogami were used as bases from which forays could be made to locate airport sites. In 1935, sites had also to be selected for radio beacons and lighting equipment. Unemployed relief personnel were used to build the installations and when Tuddy flew the new minister of transport, C.D. Howe, on an inspection tour in 1936, the Trans-Canada Airway program received a much needed boost in Ottawa.

The following year, Trans-Canada Air Lines was created, and on July 30, 1937, C.D.Howe made a historic dawn-to-dusk flight from Montreal to Vancouver in a Lockheed Electra. Stops were made at the new airfields en route for refuelling. The transcontinental airway, Howe noted, was opera-tional on schedule.

Tudhope retired from the Department of Transport to become general manager of Trans-Canada Air Lines through the Second World War. In the post-war years, he was appointed civil air attaché at Canada House, London. He donated the W.F. Tudhope Memorial Trophy to the Royal Canadian Flying Clubs Association in memory of his son William F. Tudhope, DFC,. who was killed in the Battle of Britain. The trophy was to be awarded annual-ly to the best private pilot of a flying club.

He died at the age of 65 on October 11, 1956. No man knew the terrain across Canada better.

# TURNBULL, WALLACE RUPERT

Vision coupled with a scientific outlook and an ability to 'tinker' is the essence of all pioneers of aviation. Sir George Cayley, Benjamin Franklin, and Frank Whittle were all visionaries, unafraid to experiment in the new science of aeronautics. Less well known but cast in the same mold was Canadian Wallace Turnbull.

Wallace Rupert Turnbull was born at Saint John, New Brunswick, on October 16, 1870. He graduated in mechanical engineering from Cornell University in 1893 and did post-graduate work in physics at the Universities of Berlin and Heidelberg in Germany. His first job was with the General Electric Company in New Jersey before he returned to Rothesay, New Brunswick, to work as a consulting engineer.

Always fascinated with aeronautical theory, in 1902 Turnbull built a wind tunnel to test his theories. This was the first such tunnel ever built, and Turnbull installed it in a large barn behind his home. In it he tested his theories about dihedral wing angles and airscrews. He also experimented with waterborne hydroplanes, using a two-cylinder Duryea engine for power. But Turnbull was still a theoretician, his schemes a long way from actual flight. A year later, far south of New Brunswick, the more practical Wilbur and Orville Wright made the world's first heavier-than-air flight at Kitty Hawk, North Carolina.

The first Turnbull "hydro" had a streamlined body fitted with hydroplanes beneath it. A small engine powered the two air screws set on either side of the boat in an outrigger framework. The engine was attached to the screws through a series of pulleys and belts. When started up, it tore itself to pieces before collapsing in the water.

His second "hydro" was better built and consisted of tandem floats connected to a hydrofoil on which the driver sat. The engine drove a single large propeller at the rear of the driver's seat. Turnbull had intended to fit wings onto the whole structure in the hope that the craft would lift off, but he realized that the engine was underpowered for such a feat.

W. R. Turnbull with his variable pitch propeller being tested on an Avro 504K at Camp Borden, April 1927. (NAM 2432)

With the unavailability of more powerful aircraft engines, Turnbull then turned to developing more efficient airscrews. He built a 300-foot track at the back of his home on which a small wheeled cart could be propelled back and forth. On it he mounted several versions of propellers fitted to small engines. Rudimentary instruments measured the thrust, revolutions per minute, and speed of the cart.

With the data collected, he was able to write a series of articles on the subject for scientific journals. It is chiefly through these that his work is known and remembered. The first entitled "Research on Forms and Stability of Aeroplanes" was published in *Science Review* in 1907. In it he emphasized the lack of lateral stability in aircraft of the day because of their wings. Turnbull advocated that the upward inclination of each wing from its centre towards the tip would create an angle that would better stabilize the craft.

"The Efficiency of Aerial Propellers" appeared in the same journal on April 3, 1909, and attracted much attention in both the United States and England. Two others were both titled "The Laws of Air-screws" and were published in the *Aeronautical Journal* on January 1911. For these and other research papers, the Canadian scientist received the bronze medal from the Royal Aeronautical Society and was elected a Fellow. He was now the leading authority on aerodynamics in the closely knit world of aviation.

During the First World War, Turnbull was employed by the British aircraft builders, Frederick and Company, in England. With them he continued designing propellers, one of which enabled a two-passenger plane to reach the record altitude of 13,000 feet.

The British government prevailed on him to conduct scientific experiments in other fields, one of which was the design of anti-submarine torpedo nets that could be fitted around warships.

After the war, he returned to Rothesay to work on the theory of variable-pitch propellers. He had begun experimenting with the idea in 1916 and placed a patent for it in 1922. In a controllable-pitch propeller, the blades can be adjusted at various angles or pitches during flight. This gave the pilot better control of his aircraft, especially during takeoff. In his experiments, Turnbull used a revolving electric motor and tried several types of propellers on it. In 1923, the government allowed him to test his propeller on an aircraft at Camp Borden, Ontario. These were the first-ever tests conducted on a variable pitch propeller and both the inventor and the air force were highly pleased with the results.

Five years later, Turnbull fitted a controllable-pitch propeller to an Avro biplane for a demonstration. Test pilot Flight Lieutenant G.G. Brookes took the biplane up on June 6, 1927, noting that the aircraft's performance was radically improved. The results showed that Turnbull's theories had practical benefits in the development of underpowered aircraft at that time.

The Curtiss-Wright Corporation bought the American rights to the invention, and the English rights were sold to the Bristol Aeroplane Company. Turnbull was named a Fellow to the Royal Aeronautical and the Royal Meteorological societies in England, not only for his invention of the variable-pitch propeller but his experiments with bomb sights, tidal power, and hydroplanes.But like many other inventors, his scientific curiosity satisfied, he lost interest in the invention and failed to exploit its commercial possibilities.

In 1942 the University of New Brunswick conferred an Honourary Doctor of Science degree on him, and the National Research Council in Ottawa later recognized him as a pioneer in aeronautical research in Canada. The Canadian Aeronautical Institute established the annual W. Rupert Turnbull Lectures.

Turnbull died at Saint John, New Brunswick, on November 24, 1954, having made Canada the birthplace of the variable-pitch propeller.

# VACHON, J.P. ROMEO

There is a historic old airfield in the Quebec City suburb of Ste. Foy. Today it is a park dedicated to the "Flying Postman of the North Shore." The airfield was purchased by the League of Pioneer Airmen and was given to the city to commemorate exactly where Romeo Vachon routinely took off to deliver the mail from Quebec City along the north shore of the Gulf of the St. Lawrence.

Joseph Pierre Romeo Vachon was born on June 29, 1898, at Ste. Marie de la Beauce, Quebec. He was one of the four Vachon brothers, Domat, Fernando and Irenee (Pete) — all of whom contributed to the development of aviation in Quebec. Romeo saw service in the Royal Canadian Naval Volunteer Reserve during the First World War and, in 1920, with his brother, Irenee, joined the Canadian Air Force. Both trained in Avro 504s at Camp Borden, and Romeo began his commercial pilot's career with Laurentide Air Service at Grand'Mère, Quebec, in 1922.

The Air Service had grown out of aerial forestry patrols that the Laurentide Paper Company of Quebec flew over its vast domain. The value of an aircraft in spotting forest fires was, even then, very obvious. With thousands of acres of timber at stake, pulp and paper companies began patrolling their reserves as early as 1918. War-surplus HS.2L flying boats had been acquired, and several Canadian aviation pioneers including 'Tuddy' Tudhope, Stuart Graham, and Tom Thompson gained valuable experience flying forest-fire patrols over the vast wilderness areas of Quebec and Ontario. Vachon, with his brother Irenee, flew fire-detection and some photographic flights from the company's base at Three Rivers.

Vachon's biographer, Damase Potvin, would write of the aviator's adventures:

> On the 4th of August 1923, in the Bay of Tadoussac, approximately one mile from the village, Romeo Vachon was taking off in his flying boat with his observer when it hit a

Duke Schiller on the right. Another OPAS pilot, he and Vachon rescued the crew of the *Bremen*. In 1943, while ferrying a Catalina to Britain, Schiller would disappear off Bermuda. (NAM 6097)

floating log. Under the force of the shock, the fuel tank of the plane was thrown up in the air and the two men landed in the water of the bay. The porpoises which are numerous at this time of the year surrounded them ...and eventually they reached the shore of l'Islet-aux-Morts. They walked a long distance on the round pebbles covered with slimy seaweed over which they stumbled at every step. Completely exhausted and shivering, they were spotted by Sir William Price aboard his yacht *L'Albacon* ... Sir William rescued them, putting them in discarded clothes of his crew and ... poured a bottle of rum down each of them. They recovered ... so much so that they could see thousands of porpoises all around them waiting to attack![1]

With its pilots logging 1,480 hours annually and Laurentide's fleet doubling to 12 aircraft within a year, the air service business was booming. Even when the provincial governments took over the aerial forestry business from Laurentide, the mail service more than made up for the loss. But by 1925 as improved rail and road transportation cut into the hinterland and Laurentide's revenues, the company was forced to cease business and close.

Romeo joined the Ontario Provincial Air Services until Canadian Transcontinental Airways recruited him in 1927 to fly their North Shore route. Vachon was familiar with the Gulf of the St. Lawrence from his Laurentide Air Service days. His new employers had the contract to operate a mail and passenger service from Quebec City to Seven Islands and then over to Anticosti Island. This would be extended to Rimouski and Chibougamou in 1928.

Using Fairchild monoplanes, either G-CAIP or G-CAIQ, Vachon and 'Duke' Schiller could be seen regularly taking off from the Transcontinental airfield at Ste. Foy. They tried to maintain as regular a timetable as possible. For 11 winters, with few meteorlogical or navigational aids, the Fairchilds' pilots dropped mail to isolated villages along the St. Lawrence. No parachutes were ever used. The pilot flew as low to the fields in front of the little community as possible, and the mechanic threw out the mailbags. By pioneering the dropping of mailbags and newspapers to remote villages along the desolate North Shore, Romeo Vachon acquired the name "The Flying Postman."

Outside Quebec, Vachon is best remembered for his role in the rescue of the crew of the *Bremen*. This was a W33L Junkers monoplane that would make the first east-to-west crossing of the North Atlantic Ocean in 1928. Its three-man crew had flown it from Germany to Baldonnel, Ireland, where it

Romeo Vachon, with a JN-4 Canuck at Lac à la Tortue. (NAM 2156)

left on April 12, 1928 for New York. The Junkers was blown north off its course and ran into heavy fog over Newfoundland. The pilots decided to follow whatever unknown coastline they saw beneath them, hoping it would be the United States. As they began to run out of fuel, they saw a lighthouse and crash-landed the aircraft on the rocky shore of an island. This was Greely Island, off the northwest coast of Newfoundland, and the local inhabitants ran over to meet the Junkers' crew. Word of the *Bremen*'s fate was sent by dogteam and telegraph to a waiting world, and major newspapers began chartering aircraft to rescue the crew.

Soon no less than 12 aircraft loaded with the media were making their way to Lac Ste. Agnes. One was a giant Ford Trimotor flown by the famous American aviator Floyd Bennett[2] who although very ill, had been pressed into flying a team of reporters up to Quebec.

Romeo Vachon and fellow air pioneer 'Duke' Schiller flew their Transcontinental Fairchilds through heavy snowstorms to reach Greenly Island before anyone else. Both carried not only the press but medical aid as well. Schiller and Vachon ferried the *Bremen*'s crew back, but the Junkers itself was not able to be repaired and had to be shipped in pieces to New York much later. The worldwide coverage that the rescue received made Schiller and Vachon overnight heroes.

In 1930 when Canadian Transcontinental Airways was taken over by Canadian Airways, Vachon remained to become the manager of a subsidiary airline, Quebec Airways. In 1938, when he was awarded the Trans-Canada Trophy, it was stressed that this was not for the single spectacular feat of the *Bremen* rescue, but for many years of constant flying, bringing the mail on schedule through blizzards to the isolated communities on the Gulf of the St. Lawrence.

Vachon left Canadian Airways that year to become the assistant superintendent of the Eastern Division of the new Trans-Canada Air Lines. During the Second World War, he organized aircraft overhauls for the British Commonwealth Air Training Plan. He was a member of the Air Transport Board in Ottawa until his death in 1954.

Romeo Vachon was named to Canada's Aviation Hall of fame in 1973. When Dorval Airport, Montreal, was rebuilt in the 1980s, the highway approaching it was named after Quebec's premier aviator.

---

1 Potvin, Damase, U*n hero de l'air: Heureuse aventure de Romeo Vachon* (Quebec: S.N., 1955), p. 40.
2 He had pneumonia and turned back at Murray Bay. He would be taken to Quebec City, only to die five days later.

I. 'Pete' Vachon overhauling a Vickers Viking G-CAEB, in 1926, before setting out to look for the lost gold mine. ( NAM 2171)

# VACHON, IRENEE (PETE)

Bush flying was not a job, but a way of life.[1] For the four Vachon brothers of Quebec — Romeo, Domat, Fernando and Irenee — it was the only way to live.

Irenee 'Pete' Vachon was born in Ste. Marie de Beauce, Quebec, in 1893 into what was to become an aviation-mad family. Too young to join the Royal Flying Corps and see action in the First World War, like his older brother Romeo, Pete entered the Canadian Air Force in 1920 and trained at Camp Borden. Both left the CAF in 1921 to enter commercial flying.

W.R. Maxwell was then starting Laurentide Air Services for forestry protection, and was recruiting former airforce pilots. Laurentide's aircraft were ex-U.S. Navy Curtiss HS-2L flying boats, and the operation was based at Lac à la Tortue in Quebec. The ungainly Curtiss HS-2L, described by a pilot as looking like a 'pregnant pelican,' had been designed for anti-submarine patrols, and had little cargo capacity for fire-fighting equipment. Yet they soldiered on with Laurentide and the provincial air services through the decade, providing a flying classroom for a generation of bush pilots. It was while he flew with Laurentide that Vachon acquired the nickname 'Pete.'

He was a master at engine maintenance, and years later a co-pilot recalled an incident that proved Pete's skill. Following a forced landing in the wilderness with a HS-2L, Vachon completely rebuilt the Liberty motor, scraping every bearing with only his hand tools. The engine was re-installed and operated perfectly. It was the colleague remembered later, one of the finest pieces of repair work he had ever seen.

Pete's reputation with aircraft engines would involve him in an adventure straight out of a Jack London novel. In 1926, a grizzled old prospector was making a tour of every bank in Calgary in an attempt to interest its officers in financing the search for a legendary lost gold mine in the Northwest Territories. He claimed that the map to get to it was 'in his head' so it could not be stolen. Had he not been able to show prospective backers samples of a high-grade ore, proof that he had found a rich vein of gold somewhere, the old prospector might have been rudely dismissed. He also said that, from the air, his mine wouldn't be difficult to locate as it was on the shore of a lake and that he had cut a large cross-like slash in the bush nearby. What was more, he had left his native wife behind to guard the mine.

A syndicate of financiers got enough money together to buy an aircraft and hire two pilots to fly the prospector back to his mine. It was a Vickers Viking G-CAEB from Laurentide Air Services, with a well-worn Napier engine. Jack Caldwell was engaged to fly it, and because of his experience with Napier engines, Pete Vachon was hired as the mechanic.

Just before the flight, Vachon and the prospector stopped off at a Calgary saloon for one last drink. The old man became drunk and, in the ensuing bar-room brawl, was hit on the head with a beer bottle. While this fractured his skull, it was not fatal and he recovered quickly. Unfortunately, his memory was affected — he had no recollection of a gold mine, nor a native wife, nor slashing a cross in the bush.

Undaunted, a prospecting party set out overland to the Northwest Territories, planning to meet the aircraft in the vicinity of what they thought might be the lake. Caldwell and Vachon had the Viking dismantled and trans-ported by railway to Lake La Biche, 127 miles north of Edmonton. It arrived on June 16, and Vachon set about rigging up a derrick to take the engine out and rebuild it. Parts had been sent for from England and by June 22, the air-craft was ready to be flown. That day, Caldwell and Vachon flew to Fort Fitzgerald and liaised with the main group. There, both parties proceeded north, the Viking reconnoitering ahead for suitable lakes for storing food and fuel caches.

A main base was established on a lake that the airmen named Lake Caldwell. Throughout the summer, the search for the prospector's cross went on. Flights were even made over the infamous tundra where other aircraft and canoe expeditions had disappeared before. Vachon and Caldwell were very careful in their landings and take-offs on the unknown lakes, as no one knew where they were and it only took a single crack in the hull to strand them in total isolation — and bring a lingering death by starvation. By August, the days were drawing in, and thin ice began to form on the surface of the lakes. It was only a matter of time before the hull would be damaged. While many lakes were examined, no cross slashed in the bush was ever seen. The prospec-tor's native wife must have perished from hunger as her campsite was never discovered. The search was called off and both the air and overland parties made for Edmonton.

Caldwell and Vachon were ordered to deliver the Viking to the RCAF landing strip at High River, south of Calgary. They landed there on

September 6, 1926, ending an experience that did not enrich them but caused a considerable area of the Northwest Territories to be explored. Their aircraft would end its days in a fiery crash in the Strait of Georgia in 1932. Its Napier engine that Vachon had rebuilt survived and is today preserved at the National Aviation Museum, testimony to his skill as a mechanic.

Vachon himself went on to obtain a commercial pilot's licence in 1928, and he flew for Compagnie Aerienne Franco-Canadienne. Later he would join Transcontinental Airways and finally the Curtiss-Reid Aircraft Company.

With his reputation in aeronautical engineering, he was much in demand during the Second World War and worked at Noorduyn Aviation in Montreal and later Canadian Car & Foundry. He remained in aviation until his late 70s, one of the last of the Laurentide Air Service pioneers who flew their HS-2Ls over the uncharted forests early in this century.

---

1 Bodurant, D.S., "Leaves from an Airmail Pilot's Log," *Canadian Geographic Journal*, Vol.1. No. 4, p. 332.

# WARD, Maxwell William (MAX)

✈

It is a far cry from transporting mining equipment in a biplane to a sub-Arctic outpost to becoming the chairman of the board of a multi-million dollar airline, but 'Max' Ward succeeded. He has been called the last of the true bush pilots, a visionary who took a single aircraft operation and made it a giant charter airline.

Maxwell William Ward was born in Edmonton, Alberta, on November 22, 1921. As many other young men did in 1940, he joined the RCAF hoping to fly Spitfires in the Battle of Britain, but it was not to be. In recognition of his skill as a pilot, he was chosen, like Russ Bannock and Carl Agar, to be a flight instructor in the British Commonwealth Air Training Plan, and he served at a number of stations during the course of the war.

In 1945 he began flying commercially for Northern Flights Limited from Peace River, Alberta, to Yellowknife, Northwest Territories. It wasn't long before Ward struck out on his own. On personal terms with the aviation pioneers of the previous generation, like 'Punch' Dickins and Romeo Vachon, it was little wonder that the young pilot sought to emulate them.

Using his total war gratuity, in 1946 he bought a DH Fox Moth for $10,500 and began flying prospectors and their supplies to the mining camps around Yellowknife. As his Polaris Charter Company of Yellowknife did not have an Air Transport Board Charter licence, Ward was forced into partnership with another bush pilot who possessed one. But there was a falling out and the business venture was liquidated, forcing Ward to sell his beloved Moth to pay off the debts. He learned from this and resolved that in the future he would be completely self-reliant.

Ward took to building houses to earn enough capital to start up his own company, and after much work he received a charter licence from the Air Transport Board to fly out of Yellowknife. In 1952, Wardair Ltd. began opera-

The young Max Ward with his Fox Moth at Indian Lake, NWT, 1947.
(NAM 23636)

tions with a DH Otter, it's $96,000 price financed entirely by its pilot, mechanic and owner, Max Ward.

Within the next few years, Wardair prospered. It flew mining equipment, drill crews, prospectors, and geological teams from Yellowknife to the distant edges of the Canadian North. Ward acquired a second Otter, a new Beaver, a third Otter, and in 1957 what seemed to him a mammoth aircraft — a front-loading Bristol Freighter.

Trans-Canada Air Lines had started an unsuccessful cargo service with three Bristol Freighters and sold them off one step ahead of the auditor general. With the Freighter, bulldozers, fire trucks, horses, and diesel generators could be carried by Wardair to inaccessible areas of the Yukon and Northwest Territories. In 1958, it became the first wheel-equipped aircraft to land at the North Pole and it served Wardair for a decade when the replacement of its main spar was required. Ward realised that it would be cheaper to buy a new aircraft, and this, coupled with a keen sense of history, made him donate the Freighter to the authorities at Yellowknife Airport. In full Wardair blue colour scheme, it remains today a monument to the pioneering era.

In the spirit of the legendary bush pilots who he emulated, Ward operated his fleet into remote settlements like Coppermine, Hottah Lake, Marion Lake, and Pine Point. Four former RCAF Bristol Freighters were purchased, and through the 1960s the only mishap occurred was when one fell through the ice on Great Slave Lake, NWT. It proved difficult to salvage and was stripped and sunk. The others were sold off by 1977.

In 1961, Ward changed the company's name from Wardair of Yellowknife to Wardair Canada, and he moved its headquarters to Edmonton. There were financial problems and government delays but he pressed on. In 1967, the company went public and also acquired the critical Class 9-4 International Non-Scheduled Charter licence. The passenger boom was still a few years away but Ward added a DC-6 to his fleet in anticipation of it.

The first jet aircraft, a Boeing 727 was bought in 1966, appropriately registered as C-FFUN. On its first flight, it carried vacationers from Vancouver to London, England, with a refuelling stop at Sondre Stromfjord, Greenland. Max Ward paid tribute to the bush pilots he had known by naming this and all subsequent Wardair aircraft after them. The 727 was named after Cy Becker, a Canadian air ace of the First World War, who became an early bush pilot. The aircraft did not have the range or passenger-carrying capacity to make the charter business profitable, so in 1968 a Boeing 707-320C was bought off the assembly line, with a second added the following year. Named after 'Punch' Dickins and 'Wop' May, the Boeings enabled Ward to tap the lucrative Hawaiian market.

Always aware that he was a small player in the airline business when compared to Air Canada and Canadian Pacific, Ward stressed the comfort and personal attention that neither of the other carriers could give. For example, Wardair trained their flight attendants at the PANAM school in Miami, and carried more of them in each aircraft than any other airline. Passengers praised the superb in-flight service and reliability. At a time when charter airlines were underfinanced and had dubious reputations as 'fly-by-night' concerns, Wardair pampered its passengers with real china and their own hotel accommodation packages. The Boeings could seat 183 happy vacationers, and Max Ward was well positioned to exploit the oncoming package tour boom.

Wardair entered the 1980s with Boeing 747s, (seating 452 passengers) and later DC-10s and Airbuses. The former bush airline was now flying to the Caribbean, the United States, the Orient, and the South Pacific. It obtained its first scheduled route in January 1985 and began flying scheduled services between Canadian cities and Manchester and London.

Max Ward never forgot his bush pilot origins and bought a DHC Caribou for his personal use. On July 14, 1981, while he was taking off from

Edmonton International Airport, the Caribou suffered a double engine failure and plummeted to the runway. It took all his old RCAF instructor's skills to land the powerless aircraft in a farmer's field at Morinville, Alberta. While the aircraft was being repaired, Ward bought a former RCMP single-engined Otter.[1] On summer weekends, he would fly a group of friends to his cabin in the bush. In 1985, he traded up to a Twin Otter.

In 1971, he was presented with the Billy Mitchell Award by the North American Aviation Council for his contributions to air transportation. He had a replica of his original DH Fox Moth built in 1976, but it crashed at the Toronto International Air Show the following year. It was rebuilt for display purposes and donated to the National Aviation Museum.

Seen by some as a flashy Canadian Freddie Laker, Max Ward was always keenly aware of the part that he and his airline played in the recent development of aviation in the North. To this end, he was made a member of the Order of Icarus, the Aviation Hall of Fame, and in June 1975 an Officer of the Order of Canada. His proudest moment occurred when he joined the ranks of great Canadian aviators to be awarded the Trans-Canada Trophy. It was given to him in recognition of his work in advancing air operations, not only in the Northwest Territories, but with Wardair Canada, across the world.

Although his airline would be acquired by the Pacific Western Corporation in 1989 (and the Wardair logo disappeared from the airports of the world), Max Ward is fondly recalled not only by bush pilots who saw one of their own succeed on the international scene but by millions of grateful vacationers.

---

1 Blatherwick, F.J., *A History of Airlines in Canada* (Toronto: Unitrade Press, 1989), p. 58.

# ZURAKOWSKI, JAN

L ong before the hype of having "the right stuff" or "cutting edge of technology" was bandied about, test pilots like Jan Zurakowski had made, whatever these nebulous qualities are, part of their lives. For the safety of the hundreds of pilots who came after him, this Polish immigrant would push Canadian jet fighters to the limit, working out the bugs.

Janusz Zurakowski was born on September 12, 1914, in Ryzawka, then Czarist Russia. He moved with his parents to Garwolin, Poland, and was educated at Lublin. His first flight took place as a teenager in 1929 in an ancient training aircraft. The young aviation enthusiast joined the Polish Air Force, and just prior to the war instructed at the Central Flying School in Deblin.

Jan Zurakowski, after testing the Avro Arrow for the first time. (NAM 4186)

When the German army invaded his homeland he, like hundreds of other Poles, escaped to England and flew with the Royal Air Force. Between 1940 and 1945 he fought back, flying with fighter squadrons Nos. 234, 306, 315, 316, and 609. Shot down twice, Zurakowski was credited with destroying three enemy aircraft.

By 1943 he was deputy wing leader of the Northolt Wing, and he received Poland's highest honours, the Virtuti Militari and the Polish Cross of Valour. With victory close at hand, Zurakowski was pushed toward a 'desk job' at Staff Headquarters in London. Just in time, he heard that there was an opening for a Polish pilot in the Empire Test Pilot's School and volunteered. He attended the Test Pilot's School in 1944 and for three years was a test pilot at the Aircraft and Armament Experimental establishment at Boscombe Down, England. Zurakowski had become part of that special class of aviator — the test pilot.

Retiring from the RAF as a squadron leader, he was recruited by Gloster Aircraft to fly their Meteor jet fighter. In 1950 he set a speed record in a return flight between London and Copenhagen in a Meteor 8. It was while flying it at the 1951 Farnborough Air Show that he developed the aerobatic manoeuvre called the "Zurabatic Cartwheel." That same year he flew the Gloster Javelin — known for its distinctive triangular fin and many technical problems.

Always the Polish immigrant, Zurakowski had an understated personality. Shy and unimpressive, he lacked confidence outside the cockpit of an aircraft. He was not the public's image of a dashing test pilot. At the height of his fame years later, the Toronto *Telegram* reported that he was small and balding, and looked like anything but a test pilot.

He developed his test flying techniques during the early years at Gloster. All aerobatics, however 'spontaneous' had been planned with slide rule and calculator days before. The Meteor required more than 1,500 airframe and more than 500 engine modifications, most of which had to be proven in flight.

Gloster's next aircraft the Javelin, had a delta wing and was more difficult to fly. In a dangerous series of tests, Zurakowski proved to the design office that it was unstable at a high speeds, but political reasons dictated that the aircraft was to be rushed through to operational status. The Javelin was not the success the Meteor had been because of this. The Polish pilot learned two lessons from this experience: never to take an aircraft at face value and the critical importance of collecting flight data from testing.

In 1952, he emigrated to Canada to work for the Avro Aircraft Company of Toronto. When asked later why he had chosen Canada rather than the

United States, he replied that the Canadian aviation industry had high development potential. A.V. Roe (Avro) of Toronto had just built the first jet airliner on the American continent. Closer to his heart, Avro had also unveiled the first totally Canadian jet interceptor, the all-weather CF-100. On April 21, 1952, the former Polish fighter pilot started working at the A.V. Roe plant at Malton.

On December 18, Zurakowski flew the Orenda-engined CF-100 through the sound barrier, shattering windows, it seemed, throughout Toronto suburbs. Three years later he took the CF-100 to his old stamping grounds at the Farnborough Air Show, thrilling the crowds by putting it through manoeuvres like 'the falling leaf.' No one had ever tried this with a heavy, long range interceptor before.

Called the "Clunk" or The "Lead Sled,' the CF-100 had many problems that, given time, could have been worked out, but Ottawa had committed itself by stating that the Canadian jet was now operational and ready to equip its RCAF squadrons. The RCAF itself had barely four years of flying jets with the simple Vampire, and to take on the complex "Clunks" so quickly meant that Avro's Zurakowski was constantly testing and adapting the fighter. The press were especially negative, targeting such a politically rushed, high-profile project.

Zurakowski always knew that by virtue of his role, the test pilot was an unpopular person with everyone. The designers had what he termed 'a prima donna complex.' There could not possibly be anything wrong with their design and it could not possibly require modifications. The test pilot was obviously just too fussy. The production department managers didn't like the assembly line being constantly interrupted and retooled to include these latest modifications. The head of public relations didn't like it when priority was given to an urgent technical flight and not a demonstration before a crowd of VIPs. It was little comfort for Zurakowski that only the lone RCAF pilot, far in the stratosphere, would be grateful for his meticulousness.

In 1954, the RCAF wanted heavier armament installed in the CF-100 and asked that 50 rockets be put in a special pack in the fuselage. The pack would be lowered to fire them and retracted later. Tests proved that lowering the pack produced strong vibrations and a critical change of trim. The Engineering Division sent Zurakowski and an observer up in the CF-100 to measure the stability of the aircraft with the pack lowered and retracted.

During one of these tests, at an altitude of 5,000 feet, there was an explosion in the rear of the plane and the controls became locked, forcing the aircraft into a dive. Zurakowski jettisoned the rocket pack and heard another explosion. Thinking that his observer had ejected, he did the same. When his parachute opened, he realized that he had fractured his right ankle. He landed in a field in Ajax, a Toronto suburb, and in hospital, he learned that his observer had died. The second explosion was not the ejection of the observer's seat but an explosion that damaged the escape mechanism and had incapacitated the observer.

Tests showed that the accident was caused by gasoline igniting from fractured fuel lines. These had been torn apart by the excessive vibrations when the rocket pack was being lowered. Once all the bugs had been worked out, the "Clunks" settled down to a long service career. As an all-weather interceptor, they were used in NATO and NORAD squadrons and were even exported to Belgium. Beloved in Canadian aviation folklore, the butt of RCAF jokes, they attained a particular niche in Cold War history. But not as much as Avro's next product — the "Arrow."

On March 25, 1958, at 9:51 a.m. Zurakowski flew Arrow prototype 25201 on its first test over Toronto. Although it was only a 35-minute flight, not since McCurdy's *Silver Dart* had lifted off at Baddeck, had such a milestone in Canadian aviation been reached.

No less a personage than the Minister of National Defence, George R. Pearkes, vc, stated at the Arrow 'roll-out' the year before that the manned interceptor would always be superior to the missile "because it brought the judgement of a man in the battle." Avro, already smarting from the failure of its C.102 jet airliner to attract government funding, gambled all and built the CF-105 Arrow, Mark 1. With the weight of a B-17 Flying Fortress, it could exceed Mach 2, and climb to 60,000 feet in four minutes. With its needle nose and delta wing all designed and built in less than four years, this was hailed quite rightly as a technical masterpiece — and it was totally Canadian.

Zurakowski began a strenuous flying program to evaluate the technologically advanced aircraft. He flew the second prototype in August. Speeds over 1,000 mph were reached at altitudes of over 50,000 feet. When he flew the third in September, it broke the sound barrier on its first flight! Even greater things were expected when the sixth Arrow prototype was equipped, not with Pratt & Whitney engines, but the more powerful Canadian-made Iroquois engine. With its complex weapons system designated "Astra 1," the Arrow was far superior to any aircraft that either the Russians or Americans had.

When the cost overruns of the Arrow were published, the press smelled blood and went into a feeding frenzy. Although his defence minister had spoken so warmly about the Arrow, and the RCAF wanted 150 of them, on February 23, 1958, Prime Minister John Diefenbaker decided that a missile was superior to the manned fighter and cancelled the whole project. Almost instantly, the Avro and the Canadian lead, especially in research and development, disintegrated.

The company never recovered from Ottawa's perfidity. Over 13,000 employees were laid off at Malton, all the research papers shredded and, on Ottawa's instructions, the six Arrows already built were cut up and their blueprints and photographs burned. While Canada opted for the soon-to-be obsolete BOMARC missile, both the Russians and Americans continued to operate manned bombers well into the remainder of the century.

In 1959, Jan Zurakowski was awarded the Trans-Canada Trophy in recognition of his test flying of jet aircraft. Despite this and many other honours, he remained shy and unassuming. During a reception in his honour at the Toronto City Hall, he was asked what it felt like to fly at twice the speed of sound. In a typical understatement, he replied: "It feels just like flying slowly, only faster."

But like the many who were associated with the Arrow, the test pilot was severely disappointed. Unlike his colleagues, he did not emigrate to the United States to work on the space program. The inventor of the "Zurabatic Cartwheel," the man who will always be associated with the Arrow, left Avro and aviation altogether to run a tourist resort at Barry's Bay in Ontario.